W0050413

OPTOMOTOR REFLEXES AND NYSTAGMUS

FROM THE DEPARTMENT OF OPHTHALMOLOGY IN THE WILH. GASTHUIS, AMSTERDAM
(HEAD OF THE DEPARTMENT PROFESSOR Dr A. HAGEDOORN)
AND FROM THE PRIVATE PRACTICE OF Dr G. B. J. KEINER, OPHTHALMOLOGIST, ZWOLLE

OPTOMOTOR REFLEXES AND NYSTAGMUS

NEW VIEWPOINTS ON THE ORIGIN OF NYSTAGMUS

A clinical and theoretical study of
55 cases of different forms of nystagmus in
their relationship to the optomotor reflexes

BY

Dr G. B. J. KEINER, Zwolle

AND

Dr C. OTTO ROELOFS, Amsterdam

SPRINGER-SCIENCE+BUSINESS MEDIA, B.V.
1955

This book is copyright. It may not be reproduced in whole or in part, nor may illustrations be copied for any purpose, without permission. Application with regard to copyright should be addressed to the Publisher.

ISBN 978-94-017-6700-2   ISBN 978-94-017-6764-4 (eBook)
DOI 10.1007/978-94-017-6764-4

In Memoriam Dr. G. B. J. Keiner.

Shortly before the completion of our investigation on certain forms of nystagmus it became evident that Dr. Keiner was suffering from a serious illness, and on October 2nd. 1954 his death deprived me of one of my best and truest friends and one of my most enthusiastic co-workers. Only those who were fortunate enough to be numbered among Keiner's friends will know he gave his friendship with both hands and with unequalled warmth, and will realize how distressing is the gap left by his loss. Personally, too, I shall miss him more than anyone else, as a devoted fellow-worker who found time, amid the duties of his large practice, to interest himself in scientific problems and to give his best efforts to their solution. Among his numerous cases he succeeded time and time again in discovering remarkable phenomena and in the pursuit of our investigation of nystagmus I can say with truth that it was he who continually supplied the motive force.

For me his loss as friend and co-worker is an irreparable one. I feel that the best way in which I can do honour to his memory is to promote the publication of our investigation, in the conviction that this would also have been his wish.

C. OTTO ROELOFS

CONTENTS

PREFACE

It is becoming increasingly evident that a knowledge of the physiology and pathology of the optomotor reflexes is necessary for the understanding of various functions of the organ of vision and the anomalies which may occur therein. In our clinical and theoretical study of nystagmus we have therefore not confined ourselves to the clinical picture as such but have also paid considerable attention to the theory of the optomotor reflexes. The innervation-tonus theory, which forms the basis of our considerations and several components of which have been gradually assembled by one of us (O.R.) in the course of years, is stated in full. It has developed as an extension and consequence of the theory of postural reflexes; it might be called a theory of 'Augenstellung' by analogy with that of 'Körperstellung'.

Are we breaking entirely new ground here? We do not think so. On the one hand we have Magnus and his co-workers with their research on tonic innervation and postural reflexes and on the other hand Pavlov and Zeeman, with their experimental and theoretical contributions to the study of unconditioned and conditioned reflexes, as our predecessors and guides. Nevertheless, many readers will find themselves confronted with concepts and opinions so far unknown to them. For this reason we have given in the Preliminary Notes a detailed discussion of optomotor reflexes and of some forms of nystagmus. In these considerations one is repeatedly struck by the fact that the eye, which originally was entirely under the dominating influence of the vestibular organ as far as its movements were concerned, was destined to outgrow this influence in the highest forms of life and, as a new reflexogenic centre, itself to become the source and mentor of (optical) postural reflexes.

Since the data obtained by clinical examination frequently constituted the basis of our opinions and the reliability of these data was of great importance, all the nystagmus patients were examined by two ophthalmologists, either by both the authors or by one of them with Dr. R. A. Crone (Assistant-in-Chief at

8

the University Eye Clinic, Amsterdam), whose collaboration is gratefully acknowledged.

The discussion is divided into 12 sections, each of which is more or less self-contained; this has the inevitable consequence that we are compelled to repeat ourselves from time to time. But if we have succeeded in this way in making our explanations clearer, we are willing to bear a reproach for these repetitions with equanimity.

Another reproach that may be addressed to us is that our exposition still includes many hypotheses. This is indeed the case, but still it is our opinon that these hypotheses fit together so satisfactorily that we are justified in presenting them together as a theory. The collection of data as such is undoubtedly of great value, but not until these can be combined to lead to a theory do they make rungs of the ladder up which science has to climb, even although it may appear later that some of the rungs were not altogether sound.

The Authors

A. Introduction:

The important role of the optomotor reflexes in the physiology and pathology of ocular movements has become increasingly evident as a result of recent research. The investigation described here is concerned with the great importance of these reflexes in the genesis of different types of nystagmus. Data obtained by examination of a number of patients with nystagmus have been collected and compared in an attempt to elucidate the nature and origin of some kinds of nystagmus. We restricted our investigation to those patients — 55 in number — whose nystagmus was of long standing, dating in the majority of cases from early childhood; 13 of them had pendular nystagmus and 42 had latent nystagmus.

Of the various types of nystagmus, the vestibular or labyrinthine type has probably been the most thoroughly investigated, although even here it cannot be said that all problems have been satisfactorily solved. This better knowledge of vestibular nystagmus is perhaps a reason why only a few investigators — among whom Kestenbaum (1920) must be mentioned in the first place — have paid special attention to the optomotor reflexes in their study of nystagmus. The phenomena of latent nystagmus in particular led Kestenbaum to the opinion that the optomotor reflexes were concerned to an important degree in the occurrence of this type of nystagmus.

Ohm did not share Kestenbaum's views, but in 1942 this expert on nystagmus also pointed out that latent nystagmus was still one of the most important research objects in the field of nystagmus. This remark remains equally valid now, 12 years later, in the first place because the problems of latent nystagmus have not yet by any means been solved and in the second place because their solution would undoubtedly improve our understanding of other forms of nystagmus as well and might afford us some insight into the interplay of reflexes by which the position of the eyes is regulated. We therefore considered it an advantage for our study of nystagmus that our group of patients included so many with a latent nystagmus.

One of the first points to be decided in the observation of a nystagmus is the form, i.e. whether it is a pendular or a jerking nystagmus. In most cases the distinction is not difficult, but sometimes there is uncertainty, especially if the movements are very irregular. Very often, however, a patient with pendular nystagmus will show this form only in certain directions of gaze, while in other directions it changes into a jerking nystagmus. The direction of gaze in which exclusively pendular nystagmus is present is usually that in which the eyes show the least movement ('Grundstellung' of Magnus, 1924; neutral position of Anderson 1953). This again often has the consequence that the patient holds his head in such a position that objects which catch his attention are fixated in the above-mentioned direction of gaze. One might call this a torticollis nystagmica. A similar torticollis nystagmica is also observed with a jerking nystagmus. Since the jerking nystagmus increases as a rule when the patient looks in the direction of the fast phase, this is avoided as far as possibly by turning the head in the direction of the fast phase.

In addition to the form, we note also the direction of the nystagmus movement. If the movements take place about a vertical axis, we are concerned with a horizontal nystagmus. If the movements occur about a horizontal axis lying in a plane perpendicular to the direction of gaze, we are dealing with a vertical nystagmus. If the axis is in a plane perpendicular to the direction of gaze but not horizontal or vertical, we are concerned with a diagonal nystagmus. If the axis coincides with the direction of gaze we have a rotatory nystagmus. If the axis neither coincides with the direction of gaze nor lies in a plane perpendicular to it, we are confronted with a circumductory nystagmus. This last term is seldom used, although the type of movement in question occurs frequently: it is generally referred to as a horizontal, vertical or diagonal nystagmus with a rotatory component. This terminology does perhaps define the nature of the nystagmus rather more exactly and therefore we have continued to use it.

A third important point is the amplitude or excursion of the movements and their frequency, i.e. the number of movements per unit time.

We admit freely that the recording of good nystagmograms would have been advantageous for the recognition of form, direction, amplitude and frequency and for the registration of data. Nevertheless, we did not make nystagmograms, in the first place because we did not have the opportunity to do so, secondly because we do not know of any apparatus that gives nystagmograms in which all the above features are fully brought out and thirdly because we wished to emphasize the great value of simple inspection. The human eye is an instrument which is constantly at the service of every ophthalmologist who knows how to make use of it. The tedious procedure involved in the making of nystagmograms is a deterrent to those who are not accustomed to make regular use of it. This is perhaps one of the reasons why many ophthalmologists are inclined to leave the problems of nystagmus to the otologists, neurologists and physiologists; this, however, is not right, for nystagmus also belongs to a great extent to the ophthalmologist's field and must therefore be observed, studied and judged by him. This part of his field must no longer be neglected.

B. Latent nystagmus:

The number of cases of latent nystagmus among our patients proved to be unexpectedly large. For this reason and also in view of the great theoretical importance of latent nystagmus, we wish to start with a few remarks:

In the first place, what do we understand by latent nystagmus? Briefly it may be defined as follows: Latent nystagmus is a nystagmus that first appears — or at any rate becomes enhanced to an important degree — when one of the eyes is covered or when the light-intensity or the sharpness of the image of this eye is considerably reduced. Although this definition may sound simple, it immediately gives occasion for some further remarks:

Firstly, we agree with Dupuy-Dutemps (see Bailliart 1935) that 'latent nystagmus' is a rather peculiar term. Nystagmus is not a *disease* but a *phenomenon;* either there is nystagmus or there is not; a phenomenon cannot be latent, although the affection which gives rise to it can. Nevertheless, the term 'latent nystagmus', which was introduced by A. Greafe (1880), has been so universally adopted by later authors that we do not feel justified in inventing a new name for it.

Secondly it should be noted that the clinical picture of latent nystagmus can vary greatly. We very rarely find a latent nystagmus without complications and these complications may be of many kinds: all kinds of ocular anomalies, various forms of strabismus, impairment of visual acuity and other disturbances of function.

Thirdly we should prefer to restrict the diagnosis of latent nystagmus to those patients who show a *jerking* nystagmus when one eye is covered. Thus we should not pronounce this diagnosis for patients who show when using both eyes an almost imperceptible pendular nystagmus, which increases greatly when only one eye is used. On the other hand we should speak of latent nystagmus in cases where a one-eyed patient responds to covering of the remaining eye with a jerking nystagmus having its fast component in the direction of the blind or missing eye. In such cases it would really be more correct to say that the jerking nystagmus occurs upon covering both eyes and not upon covering one of the eyes. These ideas will be more fully explained further on.

The picture of an uncomplicated latent nystagmus can be described as follows: With binocular vision the eyes appear to be perfectly still; the visual acuity is seldom greater than 6/6. When one eye is covered, a jerking nystagmus with the fast component in the direction of the uncovered eye appears. This jerking nystagmus is a coordinated movement for both eyes; it is of equal intensity with covering of either eye, but increases with abduction and decreases with adduction of the fixating eye. The decrease of visual acuity with monocular vision is due to the nystagmoid movements. When both eyes are covered, or in the dark, both eyes remain still again. In an uncomplicated case there is no strabismus.

It might here be objected, and perhaps not without justice, that this is not the picture most commonly seen. Our intention, however, was to outline the uncomplicated picture as a starting-point. The various complications will be discussed later and an attempt will be made to ascertain to what degree and in what way each of them is related to latent nystagmus.

It is not surprising that this peculiar form of nystagmus has attracted much attention. Faucon (1872), Baumeister (1873), Graefe (1880; 1898), Clarke (1896), Grimsdale (1896), Levi

(1901), Kampherstein (1903), Brewerton (1903) and Coutela (1908) were perhaps among the earliest authors to mention this picture, but to C. & H. Fromaget (1912; 1916) is due the credit for having recognized latent nystagmus as a separate syndrome and considering it as such in more detail.

Various investigators have attempted to explain the peculiar phenomena of latent nystagmus, but it cannot be said that these attempts have given a satisfactory result. Faucon (1872) and Wehrli (1916) suggested insufficiency or paresis of the external recti. Berg (1917) surmised that a reduced tonus of the external recti resulted from a decreased influx of impulses from the vestibular apparatus of the other side, i.e. a disturbance situated between the vestibular apparatus and the ocular muscles. Ohm also believed a disturbance of vestibular muscle tonus to be present, but sought its origin in the connection between retina and vestibular apparatus, but not in the centripetal part of this reflex arc because a nystagmus of this kind has never been observed to result from a local injury as such in the optic pathways, either in the optic nerve, the optic tract, the geniculate body, the optic radiation or the optic cortex (Ohm 1928). It is necessary, however, to distinguish between a developmental disturbance of a given system and an acquired lesion somewhere in the course of the optic reflex path. Dorff (1914) followed Coppez (1913) in assuming the existence of a subcortical tonic and a cortical clonic gaze centre. In latent nystagmus it is assumed that the tonic centre is disturbed, so that the clonic centre gets the upper hand. This is also held by Jacobs (1918). Hairi (1921) believes that when one eye is covered the co-ordination centre of the other side is weakened. The opinion of van der Hoeve (1917) is similar. According to this author, the primary cause is a labile equilibrium of the co-ordination centres. Since normally the left eye, when optically stimulated, exerts on the co-ordination centres an influence tending to move the eyes to the right, while the right eye when optically stimulated exerts an influence tending to move the eyes to the left, the primary labile equilibrium results in this tendency becoming manifest in monocular vision. C. Fromaget (1923) pronounced a similar opinion, except that he regarded the fast phase of the jerking nystagmus as primary, which is certainly not correct.

More important views are those of Kestenbaum (1921, 1925). While Gertz (1920) assumes the existence of an 'appareil de position' and an 'appareil de regard' and Lafon (1920) speaks

of a 'facteur statique' and a 'facteur dynamique', Kestenbaum also
distinguishes between a 'Fixationsmechanismus' and an 'Ein-
stellmechanismus'. He believes the fixation mechanism to come
into action as a result of stimuli from the fovea, each fovea
giving stimuli to ocular movements in either direction. The
fovea impulses from the right eye for deviation to the right and
those from the left eye for deviation to the left are termed homonymous reflexes; the fovea impulses from the right eye for left
deviation and those from the left eye for right deviation are
called heteronymous reflexes. Latent nystagmus is ascribed to
a disturbance of the homonymous reflexes. As we shall explain
later, we are of the opinion that the optical fixation mechanism
is not by any means confined to the fovea but that stimuli from
the periphery of the retina also play an important part in it.
The difference between the theory of van der Hoeve and that
of Kestenbaum is, thus, that v. d. Hoeve considers the primary
disturbance to be a labile equilibrium in the co-ordination centres, the predominance of the heteronymous optical fixation
reflexes being normal, whereas Kestenbaum considers that this
predominance of the heteronymous optical fixation reflexes is
not normal and that it is due to a disturbance of the homonymous optical fixation reflexes. Csapody (1923) considers the
subdivision of the eye muscle innervations into fixation reflexes
and adjustment reflexes to be artificial and is of the opinion
that in latent nystagmus the eye muscle tonus under the influence of one of the two halves of the retina predominates.

For a critical survey of the various theories we may refer to a
publication of Roelofs (1928) on latent nystagmus. Dissatis-
faction with the existing theories led him to add a hypothesis
of his own. This was related to the idea of van der Hoeve that
the primary cause was a labile equilibrium in the co-ordination
centres, and Roelofs then asked what could be the cause of this.
Since in most cases the vestibular functions were found to be
normal (Dorff 1914; Berg 1917; Lafon 1920; only Ohm repor-
ting in 1928 the observation of considerable disturbances of
vestibular reflexes in a few cases), it appeared probable that the
labile equilibrium was due to a defective development of the
tonic innervation, in so far as this is dependent on musculo-
sensory impulses and light stimuli. The assumption that a predominance of the heteronymous optical fixation reflexes was
normal was felt by Roelofs to be open to serious objections, and
he therefore endeavoured to regard this predominance as a secon-

dary phenomenon. As soon as either eye is unable, in consequence of the insufficient tonic innervation, to maintain a convergence position that it may have assumed, and deviates in a temporal direction, this will give rise reflexly to an optical stimulus to convergence movement and also to a conjugated movement in the direction of the other eye. This latter heteronymous reflex to conjugated movement will not restore the convergence position, but it will help to strenghten the tonic innervation and to promote undisturbed binocular vision. With monocular vision, however, this increased tonic innervation in the direction of the covered eye will lead to jerking nystagmus in the direction of the fixating eye.

This view was hotly disputed by Esters (1930). His opposition was based chiefly on the fact that he was unable to confirm the existence of an abnormal mode of optical localization which was observed by Roelofs in 3 patients with latent nystagmus and which served to support the latter's view. And in fact Esters was right in this; we also found out that by no means all patients with latent nystagmus show this peculiar kind of optical localization.

Esters' own ideas do not take us much further. He considers the probable cause to be a change in the ocular reflex pathways to the peripheral supranuclear relay centre and in the switching-over of these paths in that centre. Since the optical stimuli help to maintain the tonic equilibrium between the two vestibular systems and are also largely responsible for the fact that this state of equilibrium is established in the course of development, a secondary pathogenicity of the vestibular apparatus (Esters refers to the labyrinth) is not excluded. In agreement with Fromaget, Dorff, van der Hoeve, Hairi and Ohm, he is in any case of the opinion, expressed in his own way, that a lesion of the supranuclear gaze centre is involved. In opposition to Ohm's theory, which regards the vestibular nuclei as the peripheral gaze centre, he remarks, however, that covering one of the eyes has never yet been seen to cause nystagmus in any kind of affection of the labyrinth.

Thus this theory of Esters says no more than that the cause must be sought in the optic reflex pathways. It contributes little or nothing to the understanding of latent nystagmus. But the hypothesis of Roelofs also had its weak points and, as will be further set out, we were obliged to drop it. The following were among the weak points: Firstly, the so frequent occurrence of a developmental disturbance in such phylogenetically old reflexes as the musculo-sensory (proprioceptive) is rather improbable. Secondly, there is no positive evidence of a disturbance of these reflexes: where such an assumption is made it remains an *ad hoc* explanation. Thirdly, there is insufficient evidence to suggest that the predominance of the heteronymous optical fixation reflexes is really secondary and not primary.

The contradictory nature of the various theories and their many weak points, some of which have already been mentioned, made a new investigation desirable.

Anticipating the description of our experimental findings and the conclusions drawn therefrom, we may remark at this point that a primary disturbance in the development of the optomotor reflexes ultimately emerged as the most probable cause. We thus approach the ideas of Kestenbaum. On the grounds of these ideas he distinguished 2 groups of nystagmus forms; one of which he called fixation nystagmus and the other adjustment nystagmus (Kestenbaum 1921). As our views on the development and nature of the optical fixation reflexes and the optical adjustment reflexes differ from those of Kestenbaum, we shall now first deal in rather more detail with the optomotor reflexes and their development.

C. Optomotor reflexes:

Pavlov (1927) taught us to distinguish between unconditioned and conditioned reflexes. The conditioned, chiefly cortical reflexes are believed to have developed out of the unconditioned, chiefly subcortical reflexes. In every reflex we distinguish a stimulus and a reaction to that stimulus. If, now, a quite different stimulus is repeatedly applied immediately before the stimulus and reaction of an unconditioned reflex, this different stimulus will in the course of time be included as an integral part of the event and will give rise to the same reaction, so that the conditioned reflex is born.

Zeeman (1943; 1954) has worked this idea out further. Primarily he places in the centre of the picture the individual's own *action,* an action that gives rise to the impulse and to the counter-action (re-action) and thus carves out the pathway for repetition, for reproduction and amplification and for consolidation and development, as demonstrated in both phylogenesis and ontogenesis. This leads him to assume that every reflex, including the unconditioned, began as a conditioned reflex at one time in the phylogenesis or ontogenesis. The theory of neurobiotaxis (Ariens Kappers) might provide an anatomical substratum for this view.

If one consistently follows this line of thought one sees the protoplasm of the muscle cell as a source of energy; as the site of a power of action and reaction, of contraction and relaxation, that can, via centripetal nerves, make itself felt in the central nervous system, forming the first expression of the afferent part of a proprioceptive reflex arc. It is conceivable that mechanical stretching of the muscle cell forms the very first stimulus to its contraction. But as soon as the muscle cells become connected by centripetal nerves to the central nervous system, their stretching and contraction will be accompanied by a stimulation of the central nervous system and this impulse will then be propagated along centrifugal nerves and will give rise to a contraction. This is then the most primitive proprioceptive motor reflex.

It is not our intention to deal further at this point with the development of other musculo-sensory and vestibular reflexes as conditioned reflexes on the basis of these first proprioceptive reflexes. What we do propose to discuss is the manner in which the optomotor reflexes might be developed from the proprioceptive, musculo-sensory and vestibular reflexes.

Every change of position of the eye is accompanied by an increase or decrease of stretching in the various ocular muscles and the surrounding tissues. These changes in tension give rise to a proprioceptive reflex which tries to bring the eye back to its original position. Simultaneously with the displacement of the eye and the consequently changed state of stretching of the muscles, however, there occurs a displacement of images with contours over the retina. This happens every time and in this way the movement of contours over the retina becomes a conditioned reflex which is grafted onto and calibrated by the proprioceptive reflex and, like the latter, tends to maintain the original position of the eyes in the orbits. These are, of course, monocular reflexes.

In a similar way, the vestibular reflexes can also contribute to the establishment of optomotor reflexes, but in this case for conjugated movements of the eyes. With every movement of the head the eyes will execute a counter movement under the influence of vestibular stimuli, in such a way that the position of the eyes in space and the image of the external world on the retina remain more or less fixed. But with the head movement that precedes the reaction there occurs also some displacement of the image on the retina. In time this movement over the retina gives rise to a conditioned optomotor reflex which tends to keep the image of the external world on the retina fixed. This reflex will also come into action if the head movement does not occur but a displacement over the retina does occur (slow phase of the optokinetic nystagmus). This reflex is not confined to the ocular movements, and it will, guided by the vestibular reflexes, aid these reflexes in their function as placing reflexes (Magnus, 1924) and will even be able to replace them to some degree (optical placing reflexes). In this way a displacement of contours over the retinae in a given direction will, as a conditioned reflex, determine not only the position of the eyes but also the position of the head in space.

From the foregoing we may deduce that both the monocular and the conjugated optomotor reflexes serve to ensure that the retinal images do not move over the retina, both of them contributing to the optical fixation reflex. The chief part in this respect is played by the conjugated reflexes, these being grafted onto and calibrated by the vestibular reflexes, which are specially concerned with the equilibrium between the organism and the external world. The monocular reflexes will, in accordance with their origin, contribute chiefly to the primary direction of gaze. In a secondary direction of gaze they will tend rather to bring the eye back to the primary direction of gaze. In this way they constitute one of the components of the fast phase of the optokinetic nystagmus, bringing the eye back to its 'Grundstellung' (Magnus 1924) or its 'neutral position' (Anderson 1953) and thus supporting the 'Entspannungstendenz' (relaxation tendency) of Kestenbaum (1921).

In lower vertebrates and many mammals which have laterally situated eyes, so that the fields of vision have little or no overlap, there is very often a practically complete crossing of the optic nerve fibres, so that those from the right eye run to the left half of te brain and those from the left eye to the right half of the brain. [1]) Upon artificial stimulation of the end station

---

[1]) Rochon Duvigneaud (1943) points out that there is no obligate parallelism between overlapping of the 2 fields of vision and the occurrence of uncrossed fibres. The owl, for instance, has both eyes directed forwards but its nerve fibres are entirely crossed. The total absence of ocular movements in this bird is presumably significant in this connection.

in the left half of the brain in cats and dogs (Spiegel 1932) and in many other laboratory animals, both eyes turn to the right, while when the end station in the right half of te brain is stimulated both eyes turn to the left. Thus in order that the place of the images on the retina may be maintained when the contours in the field of vision shift from behind forwards, it is necessary, in accordance with what experiments with artificial stimulation have demonstrated, that in animals with complete or almost complete crossing the retina must exert an inhibitory influence on the gaze centre to which it is connected. If the displacement in the field of vision is from anterior to posterior, the retina must exert an activating (de-inhibiting) influence on the gaze centre to which it is connected.

From this it appears that an artificial stimulus cannot be regarded as equivalent to a stimulus that moves over the retina. In the latter case the stimulus remains present, although it disappears from one place at the same time that it appears at another. In this way we have to deal with a complex situation in which both activation and inhibition are at work.

Every point on the retina of such an animal, e.g. a rabbit, can thus send either an action impulse or an inhibition impulse to the gaze centre connected with the retina, quite dependently of the direction of displacement of the images over the retina. The investigation of optokinetic nystagmus has shown that impulse to movement becomes stronger the greater is the displacement of the contours over the retina. This means that the strongest activation impulses will be set up if the light stimulus reaches the most anterior part of the retina and the strongest inhibition impulses if the most posterior part of the retina is reached. From this it follows that in the course of time the stimulation of each individual part of the retina (i.e. also without movement of the stimuli over the retina) will give rise to an individually-adjusted activation or inhibition impulse by reflex means. The middle part of the retina then remains neutral; in the anterior portion of the retina the reflex action impulse increases towards the periphery and in the posterior portion of the retina the inhibition impulse increases towards the periphery. This increase of activation and inhibition is not an increase of the intensity of the impulse but of the extent of the excursion; it has thus a more qualitative character. In this way the impulses to backward and forward movement of the eyes balance each other; they maintain a *tonic innervation* which fixes the eye in the orbit as long as no displacement of images over the retina occurs.

We propose to use the term *light-tonus* for this tension pattern that forms immediately the eye receives light stimuli, although the perception of black also contributes to it by the forming of contours.

It seems probable to us that the differently graded impulses emanating from the different points on the retina constitute the physiological correlate of the optical localization.

If the objects in the field of vision move from behind forwards, and hence their images on the retina from before backwards, the action stimuli must be weakened in the anterior part of the retina and the inhibition stimuli strengthened in the posterior part. Conversely, if the objects move in the field of vision from before backwards, and hence their images from behind forwards over the retina, the inhibition stimuli must be weakened in the posterior part of the retina and the action stimuli strengthened in the anterior part.

It is a remarkable fact that ter Braak (1936) in his experiments with rabbits found that a movement from behind forwards in the field of vision constituted a stronger stimulus to following movements of the eyes than a movement from before backwards. From this one gets the impression that the transformation of an action stimulus into an inhibition stimulus produces more effect than the transformation of an inhibition stimulus into an action stimulus. This leads one to think of the induction or irradiation of inhibition (Pavlov, Lecture 9, p. 154), whereby a greater area is covered than by an action stimulus, so that the inhibition is more difficult to abolish. There are, however, so many unknowns that we do not venture to propose an explanation. It is also tempting to consider this phenomenon in connection with anomalies in the optokinetic nystagmus which may be seen in cases of cortical hemianopsia, but this also would take us too far from our actual subject.

In the higher animals and finally in man the eyes come more to the front and the accompanying semidecussation gives rise to different conditions. The temporal half of the retina becomes connected with the homolateral half of the brain; an action stimulus gives a movement in the direction of the other eye and an inhibition stimulus gives a movement in the direction of the stimulated eye. The nasal half of the retina is connected to the heterolateral side of the brain; an action stimulus gives a movement in the direction of the stimulated eye and an inhibition stimulus a movement in the direction of the other eye.

If in optokinetic stimulation an object moves from temporal to nasal in front of the eye, we get first an inhibition stimulus from the nasal half of the retina and then an action stimulus from the temporal half. If the object moves from nasal to temporal we have first an inhibition stimulus from the temporal half of the retina and then an action stimulus from the nasal half. This might be formulated as follows: *In either half of the retina a stimulus moving foveapetally (movement of contours) gives rise to an inhibition stimulus and a foveafugal contour movement gives rise to an action stimulus.* This we learn from the optokinetic nystagmus.

Now also, in the new stage of frontally situated eyes, which introduces synchronically an opposed as well as a parallel stimulation, it is still possible to find good evidence for a connection between displacement of images and retinal calibration. The frontal placing of the eyes is accompanied by the formation of a fovea, i.e. a better anatomical and functional development of that part of the retina upon which the images of objects straight in front of us are thrown. The images in the foveal region thus acquire a functional predominance and it will be chiefly the movement of retinal images away from the fovea that will give impulses to ocular movements. Thus the greatest movement away from the foveal region will give the qualitatively strongest impulse (largest excursion) to eye movements and it is understandable that this will lead to a development in which each individual part of the retina will give origin to a graded movement-impulse that will be qualitatively stronger (with greater excursion) the more peripheral is the retinal area in question.

According to this view, the eye is then no longer held in equilibrium by action and inhibition stimuli from 2 different retinal halves, but only by *action* stimuli emanating from both halves of the retina. Here again, thus, we have a tension pattern that constitutes the light-tonus and helps to fix the eye in the orbit, while this also is to be regarded as the physiological correlate of the optical localization and hence of the visual acuity. All parts of the retina thus give their own more or less accurately graded motor stimuli. The greater the visual acuity, the more accurate must this grading be; thus in the foveal region each retinal element has its own precisely graduated response. Despite these accurately graded and continually active stimuli from all the elements of the foveal region, the fovea remains motorically inactive with respect to the outside world as the resultant of the stimuli in question under normal conditions is practically zero.

From the above it follows that a foveafugal displacement of the retinal images evokes action stimuli and a foveapetal displacement evokes inhibitory stimuli, these being calibrated by in first instance proprioceptive and vestibular reflexes. The eyes are, however, subject to many other reflex influences as well, as is clearly shown by the eye movements of blind people. Owing to these various reflex influences the eyes are constantly tending to move in one direction or the other, but thanks to the resulting displacement of retinal images the tendency is in seeing people immediately suppressed and corrected by reflex means. This reflex mechanism we call the *optokinetic fixation reflex*. We wish to emphasize that this mechanism comes into action only when small displacements of the retinal images take place.

Every impulse, however small, leaves a changed stimulation condition behind it. The optical fixation reflex mechanism is based on a rapid succession of numberless motor impulses in different directions. As a consequence a higher gaze tonus is established, which we shall call *fixation tonus* to distinguish it from the light tonus described in the foregoing. The functions of the light tonus and fixation tonus are thus different. The light tonus will help to maintain the original position of the eye in the orbit and hence to maintain or restore the equilibrium with respect to one's *own body*. Its action coincides with that of the proprioceptive reflexes. The stimuli that maintain the light tonus will therefore be grafted chiefly onto the proprioceptive reflex path and will thus be of monocular nature. The fixation tonus, on the other hand, endeavours to keep the retinal image at the place where it first appeared (Kestenbaum's 'Einschnappmechanismus'). This, thus, is a maintenance of the position of the eye with respect to the *outside* world, of which the retinal image is merely a continuation. A binding of the stimuli which maintain fixation tonus to the vestibular reflexes thus appears obvious. Therefore they will be grafted chiefly onto the vestibular reflexes and will thus be of conjugated nature. Only in the neutral position is a close co-operation between the monocular and conjugated reflexes to be expected.

Optokinetic nystagmus shows us that changes in the fixation tonus can also lead to eye movements. If the optical impulses are continually given in one and the same direction, the fixation tonus is changed and the slow phase of optokinetic nystagmus occurs.

We have already remarked that the optical fixation mechanism does not come into action until displacement of the retinal image occurs; in this the fixation tonus differs from the light tonus. It is perhaps not superfluous in this connection to refer to an experiment in ter Braak's (1936) important investigation of optokinetic nystagmus in the rabbit. One of the animal's eyes was used for registration and the other eye was optokinetically stimulated. If stationary objects were also visible to the latter eye, the nystagmus did not appear as long as it was left free, but if this eye was immobilized a nystagmus was evoked in the other eye. The explanation is obvious: owing to the eye being fixed, the images of the stationary objects could not make small movements over the retina and the absence of such small movements resulted in a failure of the fixation tonus, whereby the stationary objects otherwise prevented the nystagmus, to come into action.

In fixation our eyes also continually make very small movements evoked by various subcortical and perhaps also cortical reflexes. If now the eye, in consequence of a defective development of the fixation reflexes, reacts to displacements of the retinal image in a temporal direction but not to displacements in a nasal direction, there will be, in monocular vision with that eye, a series of stimuli with successive elevation of tonus to turning of the eye in the direction of the other eye and the result will inevitably be a nystagmus which is reminiscent of optokinetic nystagmus. If this condition is present in both eyes, we have the picture of latent nystagmus.

In addition to the optical fixation reflex, which is thus identical with the reflex movement upon displacement of an image over the retina (slow phase of optokinetic nystagmus), we have also the adjusting reflex or socalled voluntary movement. Upon looking at an object, a tension pattern is formed by the combined action of the light tonus and the fixation tonus and helps to keep the eye in a fixed position in the orbit. This, however, is only the case as long as the stimuli emanating from the numerous retinal elements balance each other. If as a result of the intensity or the nature of a stimulus, or via associative pathways, a more peripheral part of the retina acquires a certain degree of predominance, the stimuli from the other parts of the retina may be inhibited and a rapid ocular movement occurs which leads to projection of the predominant stimulus on the fovea.

The inhibition of the other retinal stimuli is only of short duration; the light tonus, aided by the proprioceptive stimuli, strives to bring the eye back to its primary direction of gaze (Magnus' 'Grundstellung' — 1924). The optical fixation reflexes come into action, but for a long time impulses are still continually required to maintain the adjustment. These adjusting impulses originate from the fixated point of light and determine its optical localization. If a new fixation tonus gradually develops, the impulses no longer need to be so strong and at the same time the optical localization changes.

The convergence reaction does not differ essentially from the just mentioned adjusting reflex, and might have been developed as a conditioned reflex from the bilateral adduction innervations as unconditioned reflexes.

Thus we have taken as original, either in the phylogenetic or in the ontogenetic sense, the co-existence of 2 kinds of optomotor reflexes: (1) the monocular optomotor reflexes grafted onto lower, non-optical monocular reflex paths and (2) the conjugated optomotor reflexes grafted on lower, non-optical conjugated reflex paths. Originally, thus, either a monocular or a binocular eye movement could be evoked from each eye separately.

In everyday life a stimulus to adjusting movement of one eye will as a rule be accompanied by a stimulus to adjusting movement of the other eye. Consequently these 2 adjusting reflexes will mutually combine as unconditioned and conditioned reflexes. This physiological binding, which almost certainly has an anatomical substratum, we designate cortical binocular junction (Keiner, 1951), in contradistinction to the subcortical conjunction. We are of the opinion that convergence, as a single reflex, also requires a cortical binocular junction of this kind.

Thus, upon stimulation of one eye both a consensual movement (reflex grafted onto the conjugated reflex pathways) and an opposed movement (convergence innervation grafted onto the monocular reflex pathways) can be expected from the reflex thus evoked in the other eye. We know, however, that in a majority of circumstances the need of orientation and egocentric localization act in favour of the conjugated innervation and herewith the more practised elder sister of the convergence innervation, will predominate.

The convergence innervation requires that the cortical binocular junction shall not be a fixed point-to-point binding, as is practically the case for looking upwards and downward, but that each cortical representation of the retinal elements shall act in close relationship and therefore be connected with an area of representations of the retinal elements of the other eye ('Panumsche Empfindungskreis).

The successive overlapping of the fields of vision with more frontal placing of the eyes in the ascent of the animal kingdom, the accompanying decussation of the optic fibres and the connections which may thus be formed in the central nervous system were dealt with by Zeeman (1949) in a biologically interesting communication. Those interested are referred to this publication. It would take us too far from our subject were we to go more deeply into the matter here. On our part we may perhaps remark, in connection with the foregoing, that in the lower animals an inhibitory function, especially of the temporal half of the retina, is believed to exist and that this inhibitory function of the temporal hemiretinae from the monocular patterns after the overlapping of the fields of vision plays an important part in both convergence and fusion movements. In both cases the conjugated right and left movement is reduced to zero.

Here we intentionally make a distinction between convergence and fusion movement. Convergence innervation is based on a single impulse, which usually results from stimulation of both eyes; on account of the single nature of this impulse, double images are not produced. This convergence impulse can be regarded as the physiological correlate of depth perception. But the production of a single impulse from these stimuli coming from different eyes necessitates a cortical junction between the representations of the two retinal elements. A common single convergence innervation develops only within the range of Panum's 'Empfindungskreis'.

It is possible to imagine the convergence innervation as being built up from 2 adduction innervations which have coincided so often that in the end a unilateral stimulus suffices to give a bilateral reaction. In the case of 2 incompatible impulses to lateroversion, these will be inhibited as far as necessary. The

2 adduction impulses can then combine to give a single convergence innervation, at any rate in so far as the cortical binocular
junction permits.

If the stimuli which attract our attention impinge upon retinal
elements in either eye whose cortical representations have not
formed a cortical junction, on account of their relative positions
or for other reasons, then an impulse to adjusting movement
with inhibition of the innervation to conjugated right and left
movement will be able to produce a fusion movement, but this
is not a single convergence innervation but a double convergence
innervation and a double adduction innervation and is therefore
also accompanied by the appearance of double images. Not until
the images come within Panum's 'Empfindungskreis' is the
double adduction innervation completely taken over by a single
convergence innervation, with disappearance of the double image.
This taking-over by a convergence innervation may often be
introduced by accommodation or imagination of distance even
during the existence of the double images. Fusion, the final
coalescence of the double images, can thus only be achieved if
the stimuli from both eyes give a *common* innervation — thus
also one common convergence impulse — which is possible
only if the cortical binocular junction permits.

In connection with the foregoing we should like to draw
attention to the interesting work of Ogle (1952), whose results
fit very well into the line of thought developed by us. Ogle made
use in his investigation of 2 semi-transparent mirrors, one in
front of each eye. A fixation point was visible through the mirrors. A vertical needle was seen reflected in the mirrors. By
turning the mirrors a binocular parallax between fixation point
and needle could be brought about. When he caused the binocular
parallax to increase gradually from the beginning, 4 stages could
be distinguished in the optical localization of the needle. In the
first stage the needle was seen single but with an increasing
difference in depth with respect to the fixation point. In the
second stage the difference in depth continued to increase somewhat, but the needle was seen double. In the third stage the
distance between the double images continued to increase but
the depth difference decreased. In the fourth stage the depth
difference had entirely disappeared but the distance between
the double images, of course, continued to increase. In explaining
this phenomenon we must consider the potentiality of two
kinds of innervation impulse upon stimulation of any point on

the retina. A point lying outside the fovea has the potentiality of a consensual (homonymous) and of an opposed (heteronymous) innervation of both eyes. If the images of the needle are at different distances from the fovea in the two eyes, the unequal impulses to lateroversion may cause double images to be seen; in addition to this, however, by means of the cortical binocular junction an impulse to convergence can be sent over the original monocular reflex pathway, with the consequence of apparent depth alteration and the perception of a single image. In the first stage the difference between the impulses to lateroversion is completely abolished by inhibition. In the second stage this is no longer the case. As the apparent difference in depth still increases it is probable that the cortical representatives of the retinal points which are stimulated by the needle do after all connect up with cortical representatives of elements in the more temporal part of the retina of the other eye. In the third stage the inhibition of the lateroversion innervation — and hence the stimulus to convergence — decreases still further. In the fourth stage there is no longer any inhibition of the lateroversion innervation and the potential convergence innervation is totally inhibited.

Another remarkable fact is that if the needle is seen at a distance of some degrees peripherically from the fixation point it is still seen single at a much greater binocular parallax than is the case when its image is situated in the macular region. This phenomenon leads one to enquire whether the area of cortical junction may perhaps be larger when the stimulated part of the retina is further away from the fovea, so that this area must be assumed to lie completely in that cerebral hemisphere in which the cortical representation of the retinal point in question is situated.

D. Gaze tonus:

After the foregoing account of our view on the optomotor reflexes, in which we have made ample use of the considerations of Zeeman (1943; 1954), we now wish to say something about the gaze tonus.

The ocular muscles are perpetually in a state of contraction. This contraction state (muscle tonus) is demonstrated by the presence of muscle sounds and action currents, while it can also be inferred deductively from certain phenomena. Perhaps this muscle tonus is maintained partly by the fact that the ocular muscles are in a slightly stretched condition. In addition to this possibly autonomic muscle tonus there is certainly a reflex muscle tonus as well.

This reflex muscle tonus compels one to conclude that the ocular-muscle nuclei are in a continual state of excitation, and since the tonic innervations balance each other so accurately (we refer only to the reciprocal influences of the 2 vestibular organs) we must also assume that a reflex stimulation of the coordination centres is present.

This continual excitation state of the coordination centres, which is partly autonomic in origin but is certainly due chiefly to a constant afflux of stimuli from the periphery acting in a reflex manner, independently of our volition, is called by us the *gaze tonus*. Among the components of this gaze tonus are the light tonus, as described in the foregoing, and the variable fixation tonus developing from the fixation reflex.

The gaze tonus is subject to both quantitative and qualitative changes. A quantitative change will make the gaze tonus, and hence also the muscle tonus stronger or weaker; a qualitative change will alter the direction of gaze and hence also the reciprocal relationships of the tonus of the ocular muscles. The gaze tonus is a persisting state of excitation. It may be of non-reflex origin or be maintained by reflex means. If an after-nystagmus is observed after optokinetic stimulation, this is non-reflex, as the excitation state in the coordination centre persists autonomically although the stimuli giving rise to it have ceased. This is not the case with vestibular after-nystagmus.

The reflex gaze tonus is a more ordinary phenomenon. If, for instance, one has worn prism lenses for one day, the physiological position of rest of the eyes and hence also the excitation state of the coordination centres is altered. Although the change in the excitation state may be smaller than that following optokinetic nystagmus, it lasts very much longer. This can only be because it is reflexly maintained. The more or less voluntary adjustment of the eyes that was necessitated by the wearing of prisms was accompanied by all kinds of other stimuli (vestibular; musculo-sensory etc.) which can influence the position of the eyes. In this way we get a connection between these stimuli and the altered innervation state. The subcortical (and perhaps also cortical) reflexes have, as it were, adapted themselves as conditioned reflexes to a cortical reflex (the adjusting movement), which here has assumed to some degree the role of an unconditioned reflex.

A single innervation impulse does give a momentary change in the excitation state of the coordination centres; it interrupts the gaze tonus but does not cause any lasting change in it. Thus there is a tension between the gaze tonus and the stimuli to adjusting movement, a tension which we regard as the physiological correlate of the optical localization in the field of vision (Mesker, 1953).

A series of adjusting impulses can lead to qualitative changes in the gaze tonus. Thus the optical fixation reflex, which is built up from a series of impulses in quick succession, can very easily give rise to an altered gaze tonus. If, as in optokinetic nystagmus, these impulses are given always in the same direction, the result is a qualitative change in the gaze tonus.

A qualitative change in the gaze tonus can also occur under pathological conditions. This is clearly evident if the vestibular organ is destroyed on one side. If the right vestibular organ is destroyed, the eyes tend to deviate to the right owing to predominance of the left vestibular organ. This is opposed by other reflexes (jerking nystagmus to the left), until a new equilibrium is established by the tonic innervation of these reflexes. If now the left vestibular organ is also destroyed, the eyes tend to deviate to the left, but now not as a result of predominance of the — already destroyed — right vestibular organ but as a result of the compensatory tonic innervation that had established itself over other reflex pathways. A jerking nystagmus now also develops, but disappears again after some time as the potentiality for establishment of a new equilibrium was not yet exhausted.

If we now have two equal but antagonistic tonic innervations, no movement is to be expected as a result. It is the question, however, whether such antagonistic innervations (e.g. from both vestibular organs) cancel out and thus leave the gaze tonus unaltered, or continue to exist side by side and thus quantitatively strenghten the gaze tonus. We cannot yet give a satisfactory answer to this question. We know that in an ocular movement a relaxation of the antagonists occurs in addition to a contraction of the agonists. One might therefore surmise that two antagonistic impulses which reach the coordination centres would cancel out, provided that they are of equal intensity, and that the muscles would feel nothing of this contest. This is certainly not entirely the case, because then a quantitative increase of tonus would be excluded. One gains the impression that the action impulse for the agonists always predominates over the inhibition impulse for the antagonists and also that the activated (de-inhibited) coordination centre has more influence than the inhibited coordination centre. This would appear to apply particularly to the tonic innervation, i.e. to the reflexes responsible for the gaze tonus.

This latter seems to be in contradiction to the optokinetic nystagmus observed in the rabbit. In this animal, a movement of the objects from behind forwards in front of one of the eyes, (inhibition of the gaze centre connected to this eye) gives nystagmus more readily than movement of the objects from before backwards (activation of the gaze centre connected to this eye). It is possible that this contradiction is only apparent, as the lowering of a high gaze tonus will give more effect than the raising of an already high gaze tonus. Perhaps we must also consider the fact that an inhibition stimulus generally covers a larger area than an action stimulus (Pavlov, 1927). These relationships are somewhat reminiscent of the optokinetic nystagmus accompanying cortical hemianopsia; here also it is difficult to evoke optokinetic nystagmus with movements of objects in the direction in which the optical gaze tonus predominates.

Summarizing we may say that we find it necessary to distinguish between a fixation tonus and a light tonus as components of the optical gaze tonus. We wish to emphasize that our conception of the optical fixation reflexes and the optical fixation tonus does not entirely coincide with that given by Kestenbaum (1921, 1925) According to our ideas, the fixation reflexes can be evoked from all parts of the retina, although this may be more strongly the case in the central parts. We also think that the optical fixation mechanism does not come into action until the images start to move on the retina. In addition to the optical gaze tonus there is also a non-optical gaze tonus (vestibular; musculo-sensory; sympathetic); the significance of these two types of gaze tonus must undoubtedly vary widely in the animal kingdom.

Although our ideas on the development and nature of opto-motor reflexes and gaze tonus are still largely hypothetical, we believe that it is worth while to discuss them in some detail. In the first place we hope in this way to stimulate interest, including that of clinical ophthalmologists, as the observation of pathological conditions often provides information of value to the physiologist. Secondly we hope that the ideas presented may serve as a basis for our discussion of nystagmus and especially of latent nystagmus.

E. Optokinetic nystagmus:

One of the forms of nystagmus to which we have already referred briefly several times, and which we have shown to be of great importance for the study of optomotor reflexes and pathological nystagmus, is optokinetic nystagmus. Opinions as to the development of this physiological nystagmus are also still divided to a marked degree. In the light of what we have said about the optomotor reflexes we shall now endeavour to expound and defend our point of view with respect to optokinetic nystagmus, in order to facilitate the understanding and judgement of the conclusions we have drawn from the picture of optokinetic nystagmus in our clinical investigation. Since it was especially the horizontal optokinetic nystagmus that was of importance for our purposes, we shall confine ourselves to this.

Optokinetic nystagmus is evoked by contours which pass before the eye in a certain direction. There develops as a rule a nystagmus in which a slow and a fast phase can be distinguished. Normally the slow phase is a movement in the direction of movement of the passing contours and the fast phase is a movement in the opposite direction.

Like ter Braak (1935), we distinguish between staring nystagmus and looking nystagmus. We speak of a staring nystagmus when the patient has been instructed to stare straight in front of him, thus paying as little attention as possible to the moving contours, so that voluntary adjusting movements are avoided. In looking nystagmus, on the other hand, the attention is fixed on the moving contours; for this purpose pictures are often shown and the patient is asked to say what they represent. It is obvious that in this way the optomotor influence of the moving contours is strengthened and that of the stationary contours weakened. A danger is, however, that this may give rise to adjusting movements, i.e. to voluntary fixation of a certain part of the passing object. With the aid of the adjusting impulses a fixated following then occurs, and this is not an optokinetic nystagmus, although it is related thereto. Fixating following consists of a combination of the slow phase of optokinetic nystagmus with adjusting movements. As long as the fixated point remains visible there need be no question of an optokinetic nystagmus. If in fixating following the slow phase of optokinetic nystagmus is lacking or totally insufficient, the following movement acquires a jerky character because it consists chiefly of a series of adjusting movements.

Staring nystagmus is often regarded as equivalent to a subcortical optokinetic nystagmus and looking nystagmus as equivalent to a cortical optokinetic nystagmus. This is certainly wrong. In many decerebrated animals (ter Braak 1936) the staring nystagmus also is different in character from that in the intact animal. For man it is even very doubtful whether a subcortical optokinetic nystagmus exists at all (Velzeboer, 1952). In any case one must remember that although it may perhaps be possible to eliminate looking nystagmus and get a pure staring nystagmus, it will never be possible to produce a pure looking nystagmus without staring nystagmus.

For the further discussion of optokinetic nystagmus we propose to defend the following 3 postulates: (1) The slow phase is primary. (2) The slow phase is the consequence of an enhanced fixation tonus. (3) The fast phase is the reflex return to the intentional direction of gaze, with the aid of all systems, seperately or in combination, which are capable of contributing to this correction.

Although the great majority of authors name the jerking nystagmus according to the direction of the fast phase, most people do agree that the slow phase of optokinetic nystagmus is the primary one. Those who regard the fast phase as primary include Fromaget (1923), Ohm (1928) and Stenvers (1925). Ohm speaks of automatic rhythmic processes in Deiters' nucleus. He regards these rhythmic processes as sinusoidal vibrations and compares them with tones and overtones. It is claimed that the sinusoidal movements can be solved from a jerking nystagmus be means of a Fourier analysis. We do not doubt this possibility, but we wonder whether the sinusoidal oscillations thus found are at all representative of the physiological processes underlying the jerking nystagmus. Perhaps we can also relate this to Ohm's statements about the charging and discharging of the peripheral gaze centres. It is suggested that charging takes place during the slow phase and this is followed by a sudden discharge to the ocular muscles which produces the fast phase. An explanation of this kind is more understandable than the complicated combination of sinusoidal oscillations. If, however, no impulses are sent to the eye muscles during the charging up of the gaze centres, one is inclined to wonder how the slow phase is brought about; this would have to be a return to the resting position. Ohm knows very well that optokinetic nystagmus in animals begins with the slow phase, but he does not count the first slow phase as part of the nystagmus and speaks of a 'primäre Ablenkung' (primary deviation). But the primary deviation must also have its cause, and since it fits perfectly into the uniform acceleration shown by the slow phases in the beginning (ter Braak 1936) there is not the slightest reason to regard this first slow phase as something special.

While in animals it is easy to ascertain that optokinetic nystagmus always starts with the slow phase, this is much more difficult in man with his highly mobile eyes. There is, however, one eye movement that we cannot execute voluntarily and that is rotation about a sagittal axis. If one attempts, by rotating the contours about a sagittal axis, to evoke a rotatory nystagmus, a slow rolling of the eyes in the direction of the moving contours is the rule; there is then no sign of a precedent fast phase.

Our second postulate is that the slow phase is based on an increasing fixation tonus in one given direction. The most popular theory is that the slow phase is a following movement and the fast phase an adjusting movement. This theory would only be applicable, at the best, to looking nystagmus. If a succession of black and white stripes is used, the amplitude of the nystagmus, according to this theory, would have to be proportional to the width of the stripes, or of a few stripes together. This is frequently not the case. Totally in conflict with this theory are the facts that an optokinetic nystagmus can be evoked by a single contour, that tract hemianopsia with the boundary line passing through the fixation point does not abolish optokinetic nystagmus in either directon and that optokinetic nystagmus is not excluded even with a central scotoma.

The fact, discovered by ter Braak in 1936, that at the beginning the successive slow phases together constitute a movement with uniform acceleration is a strong support for the idea of an increasing gaze tonus. Further evidence for this is the fact that a single passing contour, which does not give rise to nystagmus, does do so if its passage before the eye is repeated a few times. Movement over a greater part of the field of vision also favours the appearance of nystagmus. Further, it is impossible to conceive of an after-nystagmus without a persistent elevation of tonus. Finally we may add that the subjective sensations (optical localization etc., see Roelofs & van der Waals, 1935) can also be the most satisfactorily explained on the basis of increasing elevation of tonus.

We need not dwell long on the origin of this rise of tonus. Our considerations on the optical fixation reflexes which come into action when images move on the retina are fully sufficient to explain the change in optical fixation tonus. As a result of this shifting of the retinal images the eyes get a continual stream of impulses to movement in order to keep the image in its place on the retina. Each innervation impulse gives only a minimal change in the tonus, which in itself would not suffice to produce the slow phase of nystagmus; the rapid succession of innervation impulses gives a change in the fixation tonus which is capable of causing the slow phase of nystagmus.

Our third postulate concerns the fast phase of optokinetic nystagmus. This has been the subject of much controversy. Bárány (1921) regarded the fast phase as an adjusting movement to a visibly approaching object. For optokinetic looking nystagmus this may occasionally be correct, but in general this view is not correct. If it were so, it would mean that no nystagmus could be evoked by a single moving contour. It would also be impossible to explain the optokinetic nystagmus in cases of total tract hemianopsia where the objects come from the direction of the blind half of the field of vision. According to Ohm the foveapetal impulses cause charging-up of the vestibular nucleus on the side from which the objects come, while the foveafugal impulses cause discharge of the vestibular nucleus on the side towards which the objects are moving. A sudden discharge of the vestibular nucleus with outflow of the impulses to the ocular muscles is regarded as giving rise to the fast phase of nystagmus. It is very strange that the moving optical stimuli should give a movement impulse (in the form of te fast phase) to the eyes which is in the opposite direction to the movement of the objects. The matter appears quite different if one imagines that the optical stimuli cause the slow phase, that the optical fixation reflexes block all those reflexes which tend to keep the eyes in their primary position and that a sudden release of this blockade causes the fast phase. An idea of this kind but rather more fully worked out has been presented by Spiegel (1932).

Ter Braak believes that the slow and fast phases are both

evoked by the same optokinetic stimulus. This opinion is based on the following observation: If the moving contours displayed to the rabbit move very slowly and the room is darkened during the slow phase, the eyes remain in the deviation position. If the light is now suddenly switched on again, the optokinetic nystagmus often begins with a fast phase under these conditions. We are inclined to think that ter Braak attaches too much importance to the fact that he moving contours become visible again and too little to the fact that the surroundings are illuminated again; the more so as he himself remarks that acoustic and tactile stimuli can then also give rise to this fast phase.

In any case, it also appears from ter Braak's experiments that optokinetic nystagmus can only begin with a fast phase if a deviation (slow phase) precedes it. This is also the opinion of Droogleever Fortuyn (1937), who further notes, from clinical observations, that the intensity of the fast phase is not governed by that of the deviation. He concludes that the fast and slow phases are bound to two separate reflexes, so that we are concerned with two contending forces which meet in one peripheral gaze centre. The slow phase is a consequence of stimuli to deviation and the fast phase is a consequence of corrective stimuli. If a deviation is produced in one way or another, the other systems will then — in combination or separately — provide the correction. As long as only one peripheral organ able to provide the corrective impulse is connected to the ocular muscle centres, the fast phase can occur. The rhythmic function of the central nervous system is then due to a rhythmic stimulation in which the stimuli may be of peripheral origin. We have adopted the view of Droogleever Fortuyn, to which we only wish to add that by deviation should be understood not a change from the primary direction of gaze but a change from the intentional direction of gaze (Kestenbaum, 1930).

The fast phase will thus occur if the deviating stimuli diminish in strength or if the correcting stimuli gain in strength. This will be the case (1) when the contours which have temporarily attracted the most attention disappear from the field of vision, (2) if as a result of increasing deviation in the slow phase the tension in the tissues and the possibly reflex 'Entspannungstendenz' (relaxation tendency) also increase and retard the slow phase and (3) if a refractory period occurs in the path of the optic fixation reflexes so that the blockade of the corrective reflexes is abolished.

The corrective reflexes may be of widely different kinds: optical, acoustic, labyrinthine, musculo-sensory or sympathetic. The optical reflexes include not only the light tonus already described but also the adjusting reflex, which will help to maintain the intentional direction of gaze. The musculo-sensory reflexes include also the proprioceptive reflexes in the restricted sense (from the eye muscles and their immediate surroundings). Some investigators (de Kleyn 1921; Stenvers 1925) have cast much doubt upon the significance of these reflexes for the fast phase. But is must not be forgotten that even if de Kleyn had succeeded in eliminating all proprioceptive stimuli without abolishing the nystagmus, this would still not prove that such stimuli do not under normal conditions contribute to the occurrence of the fast phase. Stenvers is right in saying that optokinetic nystagmus can be evoked in all directions of gaze, but this does not mean that it is the same in all directions. Further, what do we know of the relative states of tension of the contracted and uncontracted ocular muscles with different directions of gaze? Nobody, however, can deny that in deviating directions of gaze the state of tension of the ocular muscles must give a different constellation of proprioceptive reflexes from that in the primary position of the eyes.

Our considerations on optokinetic nystagmus have already given ample indications that in all kinds of disturbances in the optomotor reflexes the occurrence of disturbances in optokinetic nystagmus may also be expected. From the ideas presented it follows that these will chiefly be disturbances in the optical fixation reflexes. Since we have regularly made use of examination of the optokinetic nystagmus in order to obtain information as to the state of the optomotor reflexes, it is desirable that something should now be said about the abnormal reactions to optokinetic stimulation.

In a normal reaction to optokinetic stimulation we get a nystagmus with the slow phase in the direction of the moving contours and the fast phase in the opposite direction. The intensity of this jerking nystagmus may vary very widely, even in normal persons, so that one must not be too hasty in speaking of a pathologically intensified or reduced reaction. Judgement is particularly difficult where a pre-existent jerking nystagmus is present. If the contours move in the direction of the slow phase of the existing jerking nystagmus, then an accentuation of this nystagmus must be regarded as a normal reaction. If the contours move in the direction of the fast phase of the pre-existent jerking nystagmus, we must regard a diminution or even a total abolition of this jerking nystagmus as a normal reaction. It appears to us incorrect to speak in such cases of an inverse type or of an optokinetic insensitivity ('optische Drhestarre'). In optokinetic insensitivity the optokinetic stimulation does not bring about any change in the existing condition: immobile eyes remain immobile; a pendular nystagmus does not change its intensity nor does a jerking nystagmus become stronger or weaker. This is thus, a pathological reaction.

Another pathological reaction is the inverse type. If the eyes were at rest without optokinetic stimulation, the direction of the fast phase in an inverse type will correspond to the direction of movement of the objects and the direction of the slow phase will be opposite to it. In such cases the amplitude is nearly always small. But when are we justified in speaking of an inverse type of optokinetic nystagmus in a case of pre-existent jerking nystagmus? When the objects move in the direction of the slow phase of the pre-existent jerking nystagmus, we are entitled to speak of an inverse type if the pre-existing nystagmus becomes weakened (mild cases) or if it disappears altogether or is replaced by a jerking nystagmus with the fast phase in the direction of movement of the contours (severe cases). When the objects move in the direction of the fast phase of the pre-existent nystagmus, we may only speak of a typus inversus if this jerking nystagmus is enhanced by optokinetic stimulation.

A third form of pathological reaction is the occurrence of pendular nystagmus under the influence of optokinetic stimulation. This is seen almost exclusively with eyes having very poorly developed optomotor reflexes, so that even without optokinetic stimulation these eyes are hardly ever still.

Gradual transitions exist between these pathological reactions. We have seen patients who started with a normal but weak jerking nystagmus, followed by immobility of the eyes and finally by an inverse type. This we regard as support for the idea that the inverse type is based on an exhaustion of the optical fixation reflexes that have been called into action, so that the remaining fixation tonus in the opposite direction comes to predominate.

During the examination of optokinetic nystagmus one sometimes sees that the eyes have a certain tendency to deviate to the left or to the right. Although we know of cases in which this deviation did mean something, we still know too little about it to include it in this discussion.

F. Examination of patients:

All that now remains, before discussing our cases individually, is to outline the procedure of the examination and to explain some parts of it in more detail. After the patient's name and age had been noted, a short history was taken. Age is important because it may be assumed that secondary or compensatory phenomena will in young persons be absent or less deeply engraved than in those of riper years. In taking the history we asked whether nystagmus and possibly squint were present in other members of the family. We also endeavoured to find out at what age the nystagmus, and the squint if any, had first been noticed and to what degree these anomalies had changed during further life. If there were other ocular anomalies as well, the time of first appearance of these was also ascertained as nearly as possible.

The refraction and visual acuity were tested. We were here struck by the fact that the visual acuity was often rather poor, with binocular vision also, when the eyes appeared to be perfectly at rest.

Since Keiner (1951) found it necessary to assume the existence of a disturbance in development of the monocular optomotor reflexes in the majority of cases of squint, we invariably looked very carefully for strabismus. Great attention was paid to the possibility of alternating hyperphoria, in view of the frequency with which Crone (1952) found latent nystagmus in his 113 cases of alternating hyperphoria. It goes without saying that the necessary attention was also paid to any other ocular affection.

The nystagmus examination will be described rather in more detail. Inspection was first done with the patient looking straight ahead with both eyes open. Pendular and jerking nystagmus and their directions were first distinguished; the direction of the fast phase of jerking nystagmus was noted and the degree of liveliness of the nystagmus was ascertained. After this the nystagmus appearing upon looking to the right and to the left was described and attention was paid to changes in the nystagmus with respect to the primary direction of gaze. Particular importance was attached to the occurrence of terminal-position nystagmus, as this is indicative with a high degree of probability of a disturbance in the optical fixation reflex mechanism. As the terminal-position nystagmus lies on the borderline of the physiological, the few small adjusting jerks that often precede the definitive peripheral gaze direction must not be classed as pathological terminal-position nystagmus, nor must the practically physiological terminal-position nystagmus that most people show with protracted peripheral direction of gaze.

As with binocular vision, the nystagmus in different directions of gaze was also examined with monocular vision with each eye in turn. In many cases, but unfortunately not in all, the occurrence of nystagmus was also examined in the dark. For this purpose the patient was placed in a completely dark room and the nystagmus was ascertained by palpation, or where this was not possible by illuminating each of the eyes in turn from the side with a torch of low power; here it was always found that as far as the nystagmus was concerned it made no difference whether the right or the left eye was illuminated with this dim light.

In all cases the size of the field of gaze was ascertained for each eye separately. Only in a few patients with strabismus was this found to be slightly abnormal.

The necessary care was devoted to the testing of binocular perception. For this purpose stereoscopic pictures were used. In the first place it was ascertained whether there was also simultaneous perception of both retinal images. In a number of strabismus cases one of the retinal images was completely suppressed. Often, however, it was found possible for an object in the periphery of the visual field of one eye to be recognized simultaneously with another object somewhere in the field of vision of the other eye. The tendency to suppression appeared to be greatest for the centre of the field of vision. Several patients were capable of fusing completely or almost completely identical images in the 2 eyes, but lacked binocular depth perception. Nevertheless we also found binocular depth perception, even with difficult stereoscopic pictures, in some of our patients, although their depth perception usually took longer to appear than is the case with normal persons.

We frequently observed a remarkable phenomenon with respect to the apparent size of objects in patients with otherwise reasonably good depth perception. If in a stereoscope one displays to each eye two plane figures of equal size, the distance between the two figures being somewhat larger for one eye than for the other eye, this will give an impression of depth with binocular vision. Not only is one of the two figures seen as though it were nearer than the other but this one also appears smaller; at least this is normal. In several of our cases this apparent difference in size was not seen and there were even some patients who stated that the apparently nearer object was also apparently larger. We are of the opinion that this is connected, as is also the delay in achieving depth perception, with an insufficient development of the convergence innervation, but we do not propose to enter further into this at present.

A very striking feature in practically all cases of nystagmus, even in those with binocular perception, was the small and sometimes minimal fusion amplitude. This often made it very difficult to ascertain the existence of binocular fusion in patients without depth perception. Double images are hardly seen by such a patient; as soon as they threaten to appear, one image is suppressed. Fusion is not a reflex that can be grafted onto a reflex pathway for conjugated eye movements; thus it must have originated from the reflexes for monocular movements. We are indeed of the opinion, as we have already explained, that all oppositely directed adjusting movements have originated from the monocular movements (fusion; convergence). If the amplitude of fusion is so small, it follows logically from our conception that the development of the optical monocular reflexes must also be hampered.

Many patients were also examined with respect to a so-called normal or abnormal correspondence of the parts of the retina; this we prefer to call a normal or abnormal cortical binocular junction. For this examination we used the after-image of a vertical line of light for one eye and the after-image of a horizontal line of light for the other eye. Since the middle of the lines of light is fixated, the two after-images must form a cross if the cortical binocular junction is normal; where the cortical binocular junction is abnormal the vertical and the horizontal after-image appear more or less widely separated. In the majority of cases the two after-images were not seen simultaneously. We suspect that in such cases the cortical binocular junction has developed very poorly or not at all. Examination with the rods of Maddox also failed to give the desired results in many cases, owing to the prompt suppression of one of the retinal images.

All patients were examined for optokinetic nystagmus, both

with binocular and with monocular vision.

In a number of cases the optical localization was also studied. In some of them the optical localization corresponded to that found in the three cases described by Roelofs in 1928. But this was by no means always the case. We are of the opinion that the manner of optical localization with monocular vision in patients with latent nystagmus has something to do with the relationship between the deviating and correcting impulses, but we think it would be premature to go further into this before we have obtained more confirmation of our impression.

CASE REPORTS

In this chapter the data obtained by examination of 55 patients with nystagmus will be reported. In each case we shall endeavour to draw a general conclusion in which all the phenomena are explained as far as possible. For the sake of clarity we have arranged our cases in the following 6 groups:

Gr. I.   Patients with pendular nystagmus and practically symmetric gaze tonus.
Gr. II.   „   „   „   „   „ more or less asymmetric gaze tonus.
Gr. III.   „   „   latent   „   „ practically symmetrical gaze tonus.
Gr. IV.   „   „   „   „   „ asymmetric optical gaze tonus.
Gr. V.   „   „   „   „   „ asymmetric non-optical and
                                                    optical gaze tonus.
Gr. VI. One-eyed patients with latent nystagmus.

Gr. I. Patients with pendular nystagmus and practically symmetrical gaze tonus.

Case 1. J. W. d. B.; F., 27 yr.

Refraction [1]): both eyes M.AsM. $\left[\begin{array}{l} E + 2D \\ E + \frac{1}{2}D \end{array} \right.$ max. vert.

Vis. ac. after correction: R.E. $^1/_5$; L.E.$^1/_5$

Strab. div. oc. dextr.. No alternating hyperphoria.

Congenital pendular nystagmus had gradually disappeared. Nystagmus also stated to occur in other members of the family.

With binocular vision: to the R. and L., slight terminal-position nystagmus.

With monocular vision with R.E.: to the R., jerking nystagmus to R.; to the L., eyes at rest.

---

[1]) In the indication of the refraction we have followed Straub's method. He took the Emmetropic refraction as starting-point and indicated in this way Hypermetropia as a relatively too weak refraction with Em — x D and Myopia as a relatively too strong refraction with Em + x D.

With monocular vision with L.E.: to the R. and to the L., jerking nystagmus in the direction of gaze.

No nystagmus in the dark.

No binocular perception.

Optokinetic nystagmus:

Binocular with movement to R.: fine jerks to L.

   ,,   ,,   ,,   ,, L.: ,,   ,,   ,, R.

Monocular R.E. with movement to R.: eyes at rest.

   ,,   ,, ,,   ,,   ,,   ,, L.: jerks to R.

   ,,   L.E.   ,,   ,,   ,, R.: fine jerks to L.

   ,,   ,, ,,   ,,   ,,   ,, L.: ,,   ,,   ,, R.

The fact that the pendular nystagmus had disappeared in the course of years indicated that it could only have been very slight. This was confirmed by the existence of a practically normal although rather weak optokinetic nystagmus. There was thus, *a very slight disturbance of the development of the conjugated optomotor reflexes.* In monocular vision with the R.E. the optokinetic nystagmus was absent when the contours moved to the R. This might be due to the conjugated gaze tonus originating from the R.E. being slightly asymmetric in such a way that the gaze tonus to the L. was slightly stronger than that to the R. This assumption is supported by the observation that a terminal-position nystagmus occurred in monocular vision with the R.E. on looking to the R., whereas on looking to the L. the eyes remained at rest; apparently the resistance was overcome with more difficulty on looking to the R. than on looking to the L.

The strabismus and the very poor visual acuity in both eyes indicate that in this patient *it was chiefly the development of the monocular optomotor reflexes that was disturbed.* Strabismus divergens is to be expected if the monocular adduction reflexes are insufficient, but it can also occur if both the monocular adduction reflexes and the monocular abduction reflexes are insufficient, as the eyes will then tend to find their anatomical, divergent resting position.

Case 2. C. v. D., M., 28 yr.

Refracton: R.E. As.M. $\left[\begin{array}{l} E + {}^1/_4\,D. \\ E \end{array} \right.$;

          L.E. AsH. $\left[\begin{array}{l} E \\ E - {}^1/_2\,D. \end{array} \right.$ max. vertic.

Visual acuity with correction: R.E. $^5/_4$ with error; L.E. $^5/_4$ plus. No strabismus. No alternating hyperphoria.

Pendular nystagmus from earliest infancy. Nothing known as to heredity.

The nystagmus varied periodically in intensity; sometimes the eyes were almost still. On looking with both eyes to the R. or L. the pendular nystagmus changed to a jerking nystagmus with the fast phase in the direction of gaze (terminal-position nystagmus).

In monocular vision with the R.E.: on looking straight ahead pendular nystagmus; on looking to the R. a jerking nystagmus to the R.; on looking to the L. a very slight pendular nystagmus.

In monocular vision with the L.E.: on looking straight ahead slight pendular nystagmus; on looking to the R. a very slight jerking nystagmus to the R.; on looking to the L. a very slight jerking nystagmus to the L.

In the dark: nystagmoid movements of the eyes.

Field of gaze normal.

Good binocular perception; also depth perception. Difficult stereoscopic plates not recognized immediately. Fusion rather slow.

Optokinetic nystagmus:

Binocular with movement to R.: chiefly pendular nystagmus; sometimes jerks to L.

Binocular with movement to L.: chiefly pendular nystagmus; sometimes jerks to R.

Monocular R.E. with movement to R.: pendular nystagmus.

Monocular R.E. with movement to L.: chiefly pendular nystagmus; sometimes jerks to L.

Monocular L.E. with movement to R.: chiefly pendular nystagmus; sometimes jerks to L.

Monocular L.E. with movement to L.: chiefly jerking nystagmus to L.

As this patient had no strabismus and possessed a very good visual acuity and quite good binocular perception, it can be assumed that his monocular optomotor reflexes had developed satisfactorily. With respect to the conjugated optomotor reflexes, however, the state of affairs was quite different; the optical fixation reflexes were insufficient in all respects. This was shown by the fact that a terminal-position nystagmus almost always occurred on looking to the R. and to the L., but still more by the optokinetic nystagmus. Only very occasionally was there a trace of a normal jerking nystagmus; as a rule the pendular nystagmus persisted unchanged ('optische Drehstarre'); on many occasions a slight inverse type was observed. This inverse type can be expected only if the optical fixation tonus in the direction of the moving objects is insufficiently developed. The disturbance in this case was thus more pronounced than in Case 1. There was some evidence that in monocular vision the conjugated fixation tonus in the direction of the covered eye was somewhat less disturbed. The nystagmoid movements in the dark indicated that the non optical gaze tonus was also insufficient to keep the eyes at rest. We are of the opinion that the subcortical non-optical reflexes were insufficiently activated in consequence of the disturbance of the cortical optical reflexes.

All the phenomena can be accounted for by a *defective development of the conjugated optomotor reflexes.*

Case 3. J. V.-R.; F. 31 yr.

Refraction: both eyes emmetropic; Vis. ac. R.E. 1; L.E. 1.
No strabismus; No alternating hyperphoria.
Pendular nystagmus from the first year of life. Nothing known as to heredity.
The nystagmus had remained fairly constant; varying in intensity from time to time.
The pendular nystagmus consisted of fine, fast movements.
On looking to the R. with both eyes, jerking nystagmus to R.; on looking to the L. with both eyes, fine jerking nystagmus to L., sometimes with a rotatory component. In monocular vision with the R.E.: on looking ahead, fine pendular nystagmus; on looking to the R., jerking nystagmus to R.; on looking to the L., jerking nystagmus to L.
In monocular vision with the L.E.: on looking ahead a fine pendular nystagmus; on looking to the R. a weak jerking nystagmus to R.; on looking to the L. a somewhat stronger jerking nystagmus to the L.
Also nystagmoid movements in the dark.
Field of gaze normal. Very jerky following movements.
Good binocular perception; also depth perception; slow fusion.
Fusion amplitude $15°$. Upon exceeding the fusion amplitude, practically immediate suppression of one of the retinal images; thus no double vision.
With after-images: normal cortical binocular junction.

Optokinetic nystagmus:

Binocular with movement to R.: pendular nystagmus, stronger than with ordinary vision.

Binocular with movement to L.: pendular nystagmus, sometimes jerks to L.

Monocular R.E. with movement to R.: lively pendular nystagmus.

Monocular R.E. with movement to L.: pendular nystagmus; sometimes jerks to L.

Monocular L.E. with movement to R.: fine pendular nystagmus.

Monocular L.E. with movement to L.: pendular nystagmus.

Like the preceding patient, this patient had no strabismus, a fairly good visual acuity and reasonable binocular perception, so that any significant degree of disturbance of the development of the monocular optomotor reflexes appears unlikely.

The optical fixation reflexes were certainly quite insufficient, so that here again we can assume that there was a *disturbance in the development of the conjugated optomotor reflexes.* Upon looking to the R. or L. there always appeared a jerky terminal-position nystagmus, while optokinetic stimulation had practically no influence. The amplitude of the pendular nystagmus did, however, increase, probably because during the optokinetic stimulation the eyes lacked any point of support to which they could fix themselves by means of adjusting reflexes. On a few occasions we noted an inverse type when the objects moved to the L.; this suggests that the optical fixation reflexes for movement to the L. were still somewhat worse than those for movement to the R.

Case 4. P. J. Sch., M., 37 yr.

Refraction: R.E. M.: E. $+ 3^1/_4$ D; L.E. M.: E. $+ 3^1/_4$ D.; Vis. ac. with correction: R.E. $^1/_2$ with error; L.E. $^1/_4$.

No strabismus; no alternating hyperphoria.

Fine pendular nystagmus with now and then jerks to L.; The nystagmus was stated to have first appeared at the age of 25 yr. and to have remained constant since then. The visual acuity, however, had always been poor. Nothing known as to heredity.

On looking with both eyes to the R. or to the L. the pendular nystagmus changed into to a jerking nystagmus with the fast phase in the direction of gaze (terminal-position nystagmus).

In monocular vision with the R.E.: straight ahead, fine pendular nystagmus; to the R., a fine jerking nystagmus to the R.; to the L., a fine jerking nystagmus to the L.

In monocular vision with the L.E.: straight ahead, a fine pendular nystagmus; to the R., no nystagmus; to the L., a fine pendular nystagmus.

The nystagmus persisted in the dark.

Fields of gaze of both eyes normal.

Good binocular perception; also stereoscopic depth perception.

Small fusion amplitude. When the fusion amplitude was exceeded, double images did not appear but the image of the L.E. was immediately suppressed.

With after-images, normal cortical binocular junction.

Optokinetic nystagmus:

Binocular with movement to R.: pendular nystagmus, but stronger than with ordinary vision.

Binocular with movement to L.: pendular nystagmus, but stronger than with ordinary vision.

Monocular R.E. with movement to R.: pendular nystagmus, but stronger than with ordinary vision.

Monocular R.E. with movement to L.: pendular nystagmus, but stronger than with ordinary vision.

Monocular L.E. with movement to R.: pendular nystagmus, but stronger than with ordinary vision.

Monocular L.E. with movement to L.: pendular nystagmus, but stronger than with ordinary vision.

Nystagmus in the dark cannot, of course, be caused by optomotor reflexes. Nevertheless, congenitally blind patients also show a kind of nystagmus. From this it follows that in the absence of optical stimuli the non-optical reflexes lack an indirect influence to normal development. On these grounds we assume that in our patient the optical influence had been very slight and that the optical gaze tonus was very low. This was also shown by the terminal-position nystagmus and still more by the total absence of any reaction to optokinetic stimulation ('optische Drehstarre'). The pendular nystagmus did indeed become somewhat stronger during the optokinetic stimulation but this can be accounted for, as we have already remarked, by the fact that in consequence of the movement of the contours the eyes had no point of support which would permit them to maintain their position to some degree with the aid of adjusting reflexes. The adjusting impulses also seemed not to be strong, at any rate those of the less good (left) eye, as these did not give rise to a terminal-position nystagmus upon looking to the R. or to the L.

The absence of optical fixation reflexes points to a *rather marked disturbance in the development of the conjugated optomotor reflexes,* while in view of the insufficient visual acuity and the small fusion amplitude it is probable that the *monocular optomotor reflexes were also backward in development.*

Case 5. J.V., F., 14 yr.

Refraction: R.E. H.AsH. $\left[\begin{array}{l} E - 3^1/_2 \, D. \\ E - 4^1/_2 \, D. \end{array} \right.$;

L.E. H.AsH. $\left[\begin{array}{l} E - 4 \, D. \\ E - 4^1/_2 \, D. \end{array} \right.$

Vis. ac. with correction: R.E. $^3/_4$ with error; L.E. 1 plus.

Strabismus convergens alternans; alternating hyperphoria.

The strabismus had been noticed soon after birth; the convergence was more marked without spectacles.

Pendular nystagmus with a rotatory component from early childhood.

The nystagmus had gradually decreased.

Pendular nystagmus in all directions of gaze, both with binocular and with monocular vision.

Optokinetic nystagmus:

Binocular with movement to R.: pronouced jerking nystagmus to L.
    „   „   „   „ L.: slight jerking nystagmus to R.
Monocular R.E. with movement to R.:   good jerking nystagmus to L.
  „   „ „   „   „   „ L.:   „   „   „   „ R.
  „   L.E. „   „   „ R.: marked   „   „   „ L.
  „   „ „   „   „   „ L.: weak   „   „   „ R.

This young patient differed in some respects from the preceding patients. In the first place she had a strabismus convergens with alternating hyperphoria. This points to a *disturbance in the development of the monocular optomotor reflexes, namely those from the inferior nasal retinal quadrants* (Keiner 1951; Crone 1952). In the second place she showed a practically normal optokinetic nystagmus, which is not possible without a reasonably good development of the optical fixation reflexes. Nevertheless, since the disturbance in the monocular reflexes seemed too slight to account entirely for the pendular nystagmus, slight though this was, we are compelled to assume that there was also a *very slight disturbance in the development of the conjugated optomotor reflexes;* this applies especially to the reflexes directing the eyes to the L.

Case 6. J.U.-A., F., 29 yr.

Refraction: R.E. H.AsH. $\left[\begin{array}{l} \text{E} - 1 \text{ D.} \\ \text{E} - 2^1/_2 \text{ D.} \end{array}\right.$ max. 20° nas;

L.E. AsH. $\left[\begin{array}{l} \text{E} \\ \text{E} - 3 \text{ D.} \end{array}\right.$ max. vertic.

Vis. ac. with correction: R.E. 1 with error; L.E. $^1/_{60}$.

Strab. conv. oc. sin.; suggestion of alternating hyperphoria.

When the eyes were covered in turn the released eye showed an endorotation.

Rotatory pendular nystagmus. Other members of the family also had this affection.

With binocular vision and fixating R.E. the L.E. was somewhat higher in all directions of gaze.

In monocular vision with the R.E.: straight ahead, rotatory nystagmus with a few jerks to the R.; to the R., periodic jerks to the R.; to the L., small, irregular movements. In monocular vision with the L.E.: straight ahead and to the R., rotatory nystagmus; to the L., rotatory nystagmus with jerks to the L.

In the dark, rotatory pendular nystagmus.

Field of gaze normal.

No binocular perception.

Optokinetic nystagmus:

Binocular with movement to R.: jerking nystagmus to L.

,,   ,,   ,,   ,, L.:   ,,   ,,   ,, R.

Monocular R.E. with movement to R.: now and then jerks to L.

,,   ,, ,, ,,   ,,   ,, L.: jerking nystagmus to R.

,,   L.E. ,,   ,,   ,, R.: weak jerking nystagmus to L.

,,   ,, ,, ,,   ,,   ,, L.: eyes practically still.

This case showed a great resemblance to the previous one (Case 5). Here also there was a strabismus convergens with a suggestion of alternating hyperphoria, probably indicating a *disturbance in the development of the monocular optomotor reflexes.* The optical fixation reflexes showed certain peculiarities: With optokinetic stimulation and monocular vision the optical fixation reflexes from the amblyopic left eye appeared to be weaker than those from the good right eye. Further, with monocular vision the optokinetic nystagmus when the contours were moved in the temporal direction was weaker than when they were moved in the nasal direction. This is reminiscent of the conditions in latent nystagmus. In this case also, therefore, we conclude that there was a *slight disturbance in the development of the conjugated optomotor reflexes.*

Case 7. E.K., M., 22 yr.

Refraction: R.E. AsH. $\left[\begin{matrix} E \\ E-3\,D. \end{matrix}\right.$ max. $10°$ temp.;

L.E. AsH. $\left[\begin{matrix} E \\ E-4\,D. \end{matrix}\right.$ max. $10°$ temp.

Vis. ac. with correction: R.E. 1 plus; L.E. $^1/_{10}$ plus.

Strab. conv. oc. sin. from childhood; position correct since operation in 1947.

No alternating hyperphoria.

Rotatory pendular nystagmus.

On looking to the R. with both eyes, intensification of the nystagmus with jerks to the R.; amplitude of movements of the R.E. greater than of the L.E. On looking to the L. with both eyes, intensification of the nystagmus with jerks to the L.

In monocular vision with the R.E.: straight ahead, fine rotatory pendular nystagmus; to the R., jerking nystagmus to the R.; to the L., jerking nystagmus to the L.

In monocular vision with the L.E.: straight ahead, fine rotatory pendular nystagmus with jerks to the right each time; to the R. and to the L., terminal-position nystagmus.

No binocular perception.

Optokinetic nystagmus

Binocular with movement to R.: fine pendular nystagmus with jerks to l..
    ,,   ,,   ,,   ,, L.:   ,,   ,,   ,,   ,,   ,, ,, R.

Monocular R.E. with movement to R.: pendular nystagm.; sometimes jerks to R.

Monocular R.E. with movement to L.: fine pendular nystagmus.
    ,,   L.E. ,,   ,,   ,, R.: very slight pendular nystagmus.
    ,,   ,, ,,   ,,   ,,   ,, L.: fine pendular nystagmus.

Here also the strabismus convergens pointed to a *disturbance in the development of the monocular optomotor reflexes;* in particular it may be assumed that there was a retardation of the development of the monocular abduction reflexes, i.e. those from the 2 nasal retinal quadrants (Keiner 1951). The most prominent feature of this case, however, was a *rather marked disturbance in the development of the conjugated optomotor reflexes,* as there was practically no sign of an optical fixation reflex on optokinetic stimulation, while terminal-position nystagmus was also present on looking to the R. and to the L.

Case 8.   A., M., 27 yr.

Refraction: bilateral myopia. Vis. ac. with correction: R.E. 1; L.E. $1/3$ with error.

Strabismus convergens periodicus oc. sin.; this strabismus was stated to have developed later in life. When the R.E. was covered the correct position was maintained; when the L.E. was covered this eye was immediately adducted.

Examination with prisms showed that this convergence, or more precisely this slight unilateral adduction of the L.E. only occurred when the image of the L.E. was suppressed. This adduction with suppression of the L.E. also occurred when the patient gazed attentively at the types from a distance.

No alternating hyperphoria.

Cataracta punctata oc. sin.

Slight pendular nystagmus on looking into the distance; this disappeared with convergence at short distance. The nystagmus had first been noticed when the patient was 8 years old.

On looking to the R., a few jerky adjusting movements to the R.; on looking to the L. jerking nystagmus to the L.

In monocular vision with the R.E. the pendular nystagmus practically disappeared while the L.E. took up the position of adduction.

In monocular vision with the L.E. the pendular nystagmus persisted.

Field of gaze normal.

Good binocular perception; also stereoscopic depth perception. With rather more difficult stereoscope plates the patient needed some time to reach the correct adjustment.

The image of he left retina was very easily suppressed.

Optokinetic nystagmus

Binocular with movement to R: pendular nystagmus.

„   „   „   „ L: „   „

Monocular R.E. with movement to R.: fine pendular nyst.,
    almost still   covered L.E.

Monocular R.E. with movement to L.: fine pendular   in adduction
    nystagmus

Monocular L.E. with movement to R.:   pendular nystagmus.

„   „ „   „   „   „ L.:   „   „

The peculiar manner in which the strabismus convergens is presented in this patient must, in our opinion, be ascribed to a *disturbance in the cortical binocular junction in the right cerebral hemisphere.* We do not propose at present to go further into this mechanism, but we hope in a future publication to present our ideas on the subject illustrated by a number of case-histories. The fact that the abnormal phenomena became more marked on looking to the left and disappeared on looking to the right is probably also connected with this disturbance in the right cerebral hemisphere. Since the cortical binocular junction represents the culmination of the development of the monocular optomotor reflexes, a disturbance of it must be regarded also as a *disturbance of the development of the monocular optomotor reflexes.*

The optical fixation tonus was also insufficient, as shown by the more pronounced pendular nystagmus in monocular vision with the L.E. The terminal-position nystagmus also indicates this. The strongest evidence of insufficiency of the optical fixation reflexes, however, was provided by the optokinetic nystagmus. With optokinetic stimulation the pendular nystagmus persisted almost unchanged, so that we must conclude that there was an *important disturbance in the development of the conjugated optomotor reflexes.*

Summary of Group I.

In the 8 patients with pendular nystagmus in this group we find a marked parallelism between the intensity of the terminal-position nystagmus, the deviation from the normal optokinetic nystagmus and the intensity of the pendular nystagmus. Cases 1, 5 and 6 had little or no terminal-position nystagmus, a practically normal optokinetic nystagmus and a weak pendular nystagmus. Cases 2, 3, 4, 7 and 8, on the other hand, had a pronounced terminal-position nystagmus, practically no reaction to opto-kinetic stimulation ('optische Drehstarre') and now and then also an inverse type and a lively pendular nystagmus. This parallelism points to a common cause, in the form of an insufficiency of the optical fixation reflexes or a disturbance in the development of the conjugated optomotor reflexes. In 4 cases (Cases 5, 6, 7 and 8) strabismus was observed and in 2 of these (Cases 5 and 6) it was combined with slight alternating hyperphoria. The strabismus points to a disturbance in the development of the monocular optomotor reflexes; in Case 1 we also concluded on the grounds of the poor visual acuity that a disturbance of this kind was present. A remarkable fact is that it was just in Cases 1, 5 and 6 that only a very slight disturbance of the optical fixation reflexes was noted.

On the other hand, in Cases 2 and 3 there was no evidence of a disturbance in development of the monocular optomotor reflexes, whereas in these cases the conjugated reflexes were severely disturbed.

From these observations we conclude that a pendular nystagmus will occur especially if the optical fixation reflexes are not sufficiently capable of inhibiting involuntary movements and raising the fixation tonus, and that in some cases a defective development of the monocular reflexes may also promote the occurrence of pendular nystagmus (Cases 1, 5 and 6).

Although this group comprises only those patients who possessed a more or less symmetrical gaze tonus, some small anomalies in this respect were still noted. In Cases 3, 4, 5 and 8 the fixation tonus was more disturbed in one direction than in the other. In Cases 4 and 6 the fixation tonus from the worse eye was more disturbed than that from the better eye, while in Cases 2 and 6 with monocular vision the fixation tonus in the direction of the covered eye was somewhat less weak.

Group II. Patients with pendular nystagmus and more or less asymmetric gaze tonus.

As we have already seen, the cases of group I included a few with phenomena indicative of a slight asymmetry of the gaze tonus. These phenomena, however, were too insignificant and irregular to be worthy of much attention. Asymmetry of the optical gaze tonus can occur in 2 ways: In the first place it is possible that although the impulses from each eye do not maintain any predominant tonic innervation to left or right turning, yet the impulses from the one eye give a stronger tonic innervation than those from the other. The responsibility for this may lie with the monocular optomotor reflexes (unilateral pendular nystagmus due to insufficient light tonus) or with the conjugated optomotor reflexes (nystagmus of different intensity in monocular vision, owing to unequal fixation tonus from the 2 eyes). In the second place the asymmetry may be such that the impulses emanating from each eye give unequal tonic innervations to right and left turning. This, again, may be due to the monocular optomotor reflexes (all kinds of strabismus) or to the conjugated optomotor reflexes (latent nystagmus). Strabismus will be mentioned only cursorily in our considerations. Latent nystagmus is placed in the following groups. In this group we have placed only a few cases with pendular nystagmus and asymmetric gaze tonus.

Case 9.   W.R., M., 3 yr.

This child had a strabismus convergens oc. dextr. which had first been noticed only a few months previously. The R. eye was found to be amblyopic. Nystagmus had first been noticed when he was 10 months old. Between the ages of 12 and 14 months the nystagmus had been accompanied by concomitant movements of the head. When we examined him for the first time he showed a pendular nystagmus of the amblyopic right eye only. This nystagmus increased on looking to the right and disappeared on looking to the left; it decreased also when the R. eye was compelled to fixate. In order to exercise the R. eye and to cure its amblyopia, the L. eye was covered to exclude light for some time. This was successful in so far that the R. eye lost its nystagmus and acquired a good visual acuity. Now, however, it was found that the L. eye had become amblyopic and exhibited a pendular nystagmus which increased on looking to the left. This also responded rapidly to treatment, so that at the time of writing the child had 2 good eyes without nystagmus. His refraction was bilaterally hypermetropic: E —3 D.

The visual acuity at the time of writing was, with correction, $^5/_6$ bilaterally. The optokinetic nystagmus was normal on all occasions, both with binocular and with monocular vision, although the excursions were remarkably large and the frequency rather slow. The field of gaze was normal.

The striking correlation between amblyopia and pendular nystagmus suggests that the optical gaze tonus, in so far as this was dependent on the monocular reflexes from the amblyopic eye, was at fault. The simultaneous appearance and disappearance of amblyopia and nystagmus suggests above all a weakness and poor gradation of the foveal reflexes. The light-tonus, which serves to fix the eye in the orbit, was insufficient. This light-tonus is probably one of the components of the fast phase of the optokinetic nystagmus. This would account for the fact that in this case where the development of the optical fixation reflexes (conjugated reflexes) was further good, as shown by the practically normal optokinetic nystagmus, the frequency of this nystagmus was slow and its amplitude large.

The increase of the nystagmus in abduction suggests that the relaxation of monocular adduction innervation was not accompanied by a strengthening of the monocular abduction innervation — which was not to be expected with strabismus convergens — so that in abduction the less good eye had to depend solely on the tonus of the conjugated reflexes (fixation tonus).

When the less good eye was compelled to fixate, the nystagmus decreased; it thus appeared that fixation raised the reflex tonus.

The insufficient monocular abduction tonus had also given rise to a strabismus convergens in this patient.

We believe the phenomena observed in this case to be due to an *insufficient development of the monocular optomotor reflexes from the amblyopic eye;* this disturbance would be the more likely to manifest itself in the manner described above if the cortical binocular junction were not yet fully established.

Case 10. D.W., M., 3 yr.

Refraction: bilateral hypermetropia; E — 5½ D.

Strabismus convergens oc. sin. had been noticed shortly after birth; later strab. convergens alternans; at the time of examination, strab. conv. alternans without lenses and no strabismus with correcting lenses.

No alternating hyperphoria.

At birth, papilla grisea (Beauvieux (1926) 'pseudo-atrophie optique des nouveau-nés'), Transparent irides.

From birth the boy had shown coarse horizontal pendular nystagmus; at first not always conjugated; the amplitudes had gradually become smaller.

On looking to the R., slight terminal-position nystagmus; to the L., more marked terminal-position nystagmus.

With monocular vision the pendular nystagmus was more lively (larger amplitude) than with binocular vision.

The degree of binocular perception could not be ascertained in such a young child, but the convergence was fairly good and did not fail until a close distance was reached.

Optokinetic nystagmus

Binocular with movement to R.: pendular nystagmus.
    „   „   „   „ L.:   „   „
Monocular R.E. with movement to R.: jerking   nystagmus to R.,
    „   „ „   „   „   „ L.: pendular   „
    „   L.E.   „   „   „ R.: jerking   „   „ L.
    „   „ „   „   „   „ L.:   „   „   „ „

Apparently *the tonic innervation to left turning was less than that to right turning* in this case. This follows from he definite terminal-position nystagmus on looking to the left. In addition to this, it follows from the optokinetic nystagmus with monocular vision that for both eyes the optical fixation reflexes in the direction of the covered eye were less disturbed than those in the opposite direction; with movement of contours in the latter direction an inverse type was always present. This difference, which is also found in latent nystagmus, was greater for the left eye than for the right eye. It is therefore probable that the predominance of the gaze tonus to the right was due to stimuli from the L. eye viz.: to a deficiency of fixation tonus to the L.

Although in this child there was a *retardation of the development of the monocular optomotor reflexes* (strabismus), as is practically always the case with papilla grisea, *the most severe disturbance was in the development of the conjugated optomotor reflexes* ('optische Drehstarre' and inverse type with optokinetic stimulation).

Case 11. E.R., F., 29 yr.

Refraction: R.E. MAsM. $\left[\begin{array}{l} E + 3\,D. \\ E + 7\,D. \end{array} \right.$ max. 85° temp.;

L.E. MAsM. $\left[\begin{array}{l} E + 2\,D. \\ E + 5\,D. \end{array} \right.$ max. 85° temp.

Vis. ac. with correction: R.E. $^3/_5$; L.E. $^3/_5$.

Trace of alternating hyperphoria. After covering and uncovering, each eye came from somewhat nasally above.

Albinotic fundi; transparent irides and sclerae.

The patient's twin sister was also albinotic and a brother had nystagmus.

Pendular nystagmus with jerks to R.

On looking with both eyes to the R., jerking nystagmus to R.; to the L., weak jerking nystagmus to L.

In monocular vision with the R.E.: straight ahead, pendular nystagmus with jerks to R.; to the R., jerking nystagmus to R., to the L., weak jerking nystagmus to L.

In monocular vision with the L.E.: staight ahead, very slight jerking nystagmus to R.; to the R., pronounced jerking nystagmus to R.; to the L., very weak jerking nystagmus to L.

In the dark, pendular nystagmus with a few jerks to R.

Field of gaze normal.

Good binocular perception; also depth perception. Fusion amplitude approx. $25°$.

Optokinetic nystagmus

Binocular with movement to R.: pendular nystagmus with jerks to L.
        ,,   ,,   ,,   ,, L.:   ,,   ,,   ,,   ,, ,, R.

Monocular R.E. with movement to R.: pendular nystagmus with jerks to R.
  ,,   ,, ,,   ,,   ,,   ,, L.:   ,,   ,,   .
  ,,   L.E. ,,   ,,   ,, R.:   ,,   ,, with jerks to R.
  ,,   ,, ,,   ,,   ,,   ,, L.:   ,,   ,,   .

The nystagmus with jerks to R., the stronger terminal-position nystagmus on looking to the R. and the optokinetic nystagmus with monocular vision show clearly that *the fixation tonus to the R. was weaker than that to the L.* The jerking nystagmus to R. with monocular vision with the L.E. was stronger than with monocular vision with te R.E. On looking to the L. with the L.E. the jerking nystagmus to the L. did not occur until later and was weaker than on looking to the L. with the R.E. This indicates that especially for the L.E. the fixation tonus to the R. was weaker than that to the L. This conflicts with what is usually observed with latent nystagmus. Strangely enough, Ohm (1930) also described an albinotic patient with latent nystagmus in whom the jerks with monocular vision were in the nasal direction (See also Cases 12 and 13).

In any case there was, in our patient, a *rather marked disturbance of the development of the conjugated optomotor reflexes,* while the slight alternating hyperphoria and the insufficient visual acuity indicate that the *development of the monocular optomotor reflexes was also not quite normal.*

Case 12. R.P., M., 5 yr.

Visual acuity: R.E. $^3/_5$; L.E. $^2/_5$.

Wandering, dissociated eye movements had been noticed shortly after birth; the pupils reacted to light and darkness. At the age of $8^1/_2$ months the eye movements were completely associated and a good convergence was present.

No strabismus. No alternating hyperphoria.

Partial albinism.

Irregular, fine pendular nystagmus alternating with periods of rest.

With distant vision, pendular nystagmus appeared every time; with close vision the irregular pendular nystagmus occurred less frequently.

On looking to the R. and to the L., marked terminal-position nystagmus with both binocular and monocular vision. The pendular nystagmus was more marked with monocular than with binocular vision.

The convergence reaction with monocular vision with the R.E. was better than that with monocular vision with the L.E.

Optokinetic nystagmus

Binocular with movement to R.: rather irregular jerking nystagmus to L.
   ,,   ,,   ,,   ,, L.: pendular nystagmus with jerks to L.
Monocular R.E. with movement to R.: irregular jerking nystagmus to L.
   ,,   ,, ,, ,,   ,,   ,, L.:   ,,   movement with jerks to L.
   ,,   L.E. ,,   ,,   ,, R.:   ,,   ,,   ,,   ,,   ,, L.
   ,,   ,, ,, ,,   ,,   ,, L.:   ,,   ,,   ,,   ,,   ,, R.

As for the preceding patient, we are inclined to relate this child's nystagmus to his partial albinism, as a result of which less sharply graded retinal stimuli were given. The insufficient visual acuity might then point to a *disturbance in the development of the monocular optomotor reflexes* and the poor reaction to optokinetic stimulation, as well as the terminal-position nystagmus, might point to a *rather marked disturbance in the development of the conjugated reflexes.* The optokinetic nystagmus, both in binocular vision and in monocular vision with the better R. eye indicate a *predominance of the tonic inner-*

vation to right turning from the R. eye. The tonic innervation from the L. eye, although presumably very low, was better balanced.

A predominance of the tonic innervation in the temporal direction in monocular vision, as here observed for the R. eye, was found for the left eye in the previous patient. This is rather rare; we draw attention once more to the fact that both these patients were albinotic, as was also Ohm's patient.

Case 13. M.L., F., 13 yr.

Refraction: R.E. ?; L.E. Hm. E — $4^{1}/_{2}$ D.; Vis. ac. R.E.$^{1}/∞$; L.E. with corr. $^{1}/_{6}$. Congenital aniridia of both eyes; cataracta calcarea oc. dextr.

Pendular nystagmus; eyes sometimes at rest for a short time; sometimes wandering movements; occasionally jerking nystagmus to the R. The eye movements were reminiscent of the nystagmus of blind persons.

On looking to the R., jerking nystagmus to the R. every time; on looking to the L. no jerking nystagmus but only pendular nystagmus.

The pendular nystagmus also persisted in the dark.

Optokinetic nystagmus

with movement to R.: jerking nystagmus to R. with small amplitude.
„ „ „ L.: „ „ „ L. „ „ „

Covering of the practically blind R. eye had no influence. No optomotor reflexes could be evoked from this eye.

The optomotor reflexes from the less bad L. eye were also poorly developed. This was evident from the wandering eye movements seen from time to time and from the inverse type of reaction to the optokinetic stimuli. The jerking nystagmus to R. on looking to the R. and the absence of a jerking nystagmus to L. on looking to the L. make it probable that the tonic innervation to left turning predominated somewhat over that to right turning. Here again we have thus a *predominance of the temporally-directed tonic innervation in monocular vision.* Although in this case we are not concerned with albinism as in the 2 previous cases, this was a patient with a very serious ocular defect which can be held responsible for the poor development of the optomotor reflexes. This leads to the conjecture that the possibility of a predominance of the temporally-directed fixation tonus may be considered more likely if ocular defects have impeded the development of the optomotor reflexes, whereas a predominance of the nasally-directed fixation tonus may be expected more in those cases where the interference with development of the reflexes has been due to more central causes. It might be asked whether the predominance of the gaze tonus in the temporal direction cannot perhaps be ascribed to non-optical reflexes. This, however, appears extremely unlikely in view of the fact that we have never seen any patient with a predominance of the non-optical gaze tonus in a given direction, in whom an inverse type appeared with optokinetic stimulation when the contours passed before the eye in the direction in question. We therefore adhere to our opinion that here also we have a case in which *both the monocular and the conjugated optomotor reflexes had been unable to develop* on account of ocular anomalies.

Summary of Group II.

The 5 patients with pendular nystagmus who made up this group showed an asymmetry in the gaze tonus that leads on closer analysis to important conclusions. In Case 9 the appearance of pendular nystagmus ran parallel to that of amblyopia. This indicated that a disturbance of development of the monocular optomotor reflexes can also cause a pendular nystagmus. Since the optokinetic nystagmus was normal and there was no terminal-position nystagmus, it could hardly be assumed that there was any disturbance of the development of the conjugated optomotor reflexes. It might be asked why patients with strabismus do not more frequently show a nystagmus of the amblyopic eye. The answer is probably that amblyopia only indicates a disturbance of the monocular optomotor reflexes from the centre of the retina and that pendular nystagmus will appear only if the monocular optomotor reflexes (i.e. the light-tonus) from the whole of the retina are at fault.

Case 10 showed, in addition to pendular nystagmus a suggestion of predominance of the nasally-directed gaze tonus with monocular vision, this being reminiscent of latent nystagmus. Cases 11, 12 and 13, on the other hand, showed a predominance of the temporally directed gaze tonus in monocular vision, although only with one eye but not the other. The question arises of whether this temporally-directed predominance may be expected just in those cases in which ocular anomalies have interfered with the normal development of the optokinetic reflexes.

Group III. Patients with latent nystagmus and practically symmetric gaze tonus:

This group comprises those patients with latent nystagmus whose optical gaze tonus evoked by retinal stimuli to the R.E. was practically equal but opposite to that evoked by stimuli to the L.E. The resultant, thus, when both eyes are open, is a symmetrical gaze tonus that tends to direct the eyes straight ahead. In this group the non-optical stimuli also maintained a practically symmetrical gaze tonus. Of the 35 cases of latent nystagmus examined by us in the last few years, 15 conformed to the above definition. Their histories follow, being arranged as far as possible in ascending order of severity of the affection.

Case 14. H.O., M., 17 yr.

Refraction: bilat. hypermetropia $E - 2^1/_2$ D.
Vis. ac. with correction: R.E. $^3/_5$; L.O. $^3/_5$; binoc. $^5/_5$.
No strabismus; no alternating hyperphoria.
Latent nystagmus.
On looking with both eyes to R. or L. often some dissociation, either convergent or divergent.
Field of gaze normal, although there was some difficulty in glancing far to the side.
Binocular perception with good convergence but small fusion amplitude.

Optokinetic nystagmus

Binocular with movement to R.: jerking nystagmus to L.
     „   „   „   „ L.:   „   „   „ R.
Monocular R.E. with movement to R.: jerking nystagmus to R.
  „   „ „   „   „ L.:   „   „   „ „
Monocular L.E. with movement to R.: jerking nystagmus to L. with large amplitude.
Monocular L.E. with movement to L.: jerking nystagmus to L. with large amplitude.

This case represented one of the most *uncomplicated* forms of latent nystagmus. The less good visual acuity in monocular vision must be ascribed to the movement of the eyes, as the acuity became much better with binocular vision. There was no evidence of a disturbance in the development of the monocular optomotor reflexes, but there was a *disturbance in the development of the conjugated optomotor reflexes,* the most affected being the reflex which, evoked in either eye, served to move this eye in the direction of the stimulated eye. With binocular vision the optokinetic nystagmus was practically normal; with monocular vision the latent nystagmus took complete command.

Case 15. K. Sch., M., 7½ yr.

Refraction: bilat. hypermetropia E — 2 D.
Vis. ac. with correction: R.E. $^5/_5$; L.E. $^5/_5$ (formerly $^2/_5$); binoc. $^5/_5$.

Strabismus convergens periodicus oc. sin. since the age of 2 yr. Formerly also with sursumvergent tendency but this had now disappeared; no alternating hyperphoria.

Latent nystagmus.

Nothing known as to hereditary or familial occurrence of latent nystagmus.

In binocular vision the eyes kept still, also on looking to the R. or L. In monocular vision there was only a slight jerking nystagmus to temporal which increased somewhat with looking in the temporal direction and disappeared with looking towards the nose.

No nystagmus in the dark.

Field of gaze normal.

Binocular perception and depth perception were present, but ability to estimate sizes was unsatisfactory. By this we mean that the apparent sizes of objects did not decrease with increasing covergence.

Fusion amplitude only 9°.

Optokinetic nystagmus

Binocular with movement to R.: jerking nystagmus to L.
,,   ,,   ,,   ,, L.:   ,,   ,,   ,, R.
Monocular R.E. with movement to R.: weak jerking nystagmus to L.
,,   ,, ,,   ,,   ,,   ,, L.: jerking nystagmus to R. with large amplitude.
Monocular L.E. with movement to R.: jerking nystagmus to L. with large amplitude.
Monocular L.E. with movement to L.: weak jerking nystagmus to R.

This was a case of very slight and practically uncomplicated latent nystagmus. The strabismus convergens periodicus and the small fusion amplitude showed, however, that in addition to the *slight disturbance in the development of the conjugated opto-motor reflexes* (slight latent nystagmus), there must also be a *slight disturbance in the development of the monocular opto-motor reflexes*. That the disturbance in the conjugated reflexes was not pronounced was shown by the fact, among other things, of the change in the nystagmus on optokinetic stimulation with monocular vision. With movement in the nasal direction the existing nystagmus was accentuated and with movement in the temporal direction we saw a jerking nystagmus in the nasal direction. The optical fixation reflexes were capable of over-coming the latent nystagmus.

Case 16. G.D., F., 9 yr.

Refraction: bilat. myopia E + 7D.
Vis. ac. with correction: R.E. $^4/_5$; L.E. $^5/_5$; binoc. $^5/_5$.
Strab. conv. periodicus oc. dextr. (15 — 20°); stated to have been present from birth.
With spectacles the eyes remained in the parallel position most of the time.
No alternating hyperphoria.
The child was rather backward; she showed bilateral epi-canthus.
Latent nystagmus; when the L.E. was covered the amplitude was somewhat less than with the R.E. covered.

Optokinetic nystagmus

Binocular with movement to R.: very weak jerking nystagmus to L.
„ „ „ „ L.: „ „ „ „ „ R.
Monocular R.E. with movement to R.: some irregular movements of the eyes.
Monocular R.E. with movement to L.: jerking nystagmus to R.
„ L.E. „ „ „ R.: „ „ „ L.
„ „ „ „ „ „ L.: only a few jerks to L.

This child's latent nystagmus was more lively than that of Case 15.
The disturbance in the development of the conjugated opto-motor reflexes was thus probably somewhat greater. This sur-mise was confirmed by the reaction to optokinetic stimulation.

Comparison of this with that of Case 15 shows that although optokinetic stimulation was not entirely without influence, it had less effect than in Case 15. The strabismus indicates that there was also a *disturbance in the development of the monocular optomotor reflexes.*

Case 17. W.G., F., 13 yr.

Refraction: bilat. hypermetropia $E - 2^1/_2$ D.
Vis. ac. with correction: R.E. 1 with error; L.E. 1; binoc. 1.
Strabismus convergens sursumvergens alternans periodicus. A squint had been present from birth; formerly it had been permanent and chiefly with the R.E.
Alternating hyperphoria.
Variable latent nystagmus.
In binocular vision: straight ahead, sometimes slight pendular nystagmus; slight terminal-position nystagmus on looking to R. or L.
In monocular vision with R.E.: slight jerking nystagmus to R. and higher position of the L.E.
In monocular vision with L.E.: slight jerking nystagmus to L. and marked adduction of the R.E., which also turned slightly upwards.
Convergence reasonable but no binocular depth perception.

Optokinetic nystagmus

Binocular with movement to R.: Often no nystagmus; with the L.E. fixating, somet'mes jerks to L.
Binocular with movement to L.: suggestion of jerking nystagmus to R.
Monocular R.E. with movement to R.: fine jerks to R.

" " " " " " L.: jerking nystagmus to R.
" L.E. " " " R.: " " " L.
" " " " " " L.: vibratory movements of the eyes or fine jerking nystagmus to L.

As in Case 16 we see that optokinetic stimulation has an unmistakable effect on the nystagmus with monocular vision but cannot overcome it completely. Here, thus, we are again concerned with a *moderately severe disturbance of the development of the conjugated optomotor reflexes.* Nevertheless, this case differed in certain respects from the preceding one: In the first place by the possession of a clear-cut alternating hyperphoria.

According to Crone (1952), whose opinion we readily endorse, such an alternating hyperphoria is attributable to a disturbance in the monocular optomotor reflexes originating from the lower nasal retinal quadrants. In the second place, Case 17 showed the peculiar phenomenon that when the R.E. was covered (sometimes even with fixation by the L.E.) the covered eye took up a position of pronounced adduction. This adduction was not present in the dark; it was thus a consequence of light stimuli received by the L.E. Something similar was also seen in Case 8. We ascribe anomalies of this kind to a disturbance in the cortical binocular junction, a subject with which we hope to deal in more detail in a future publication. In any case, we must conclude that this patient also had a *disturbance in the development of the monocular optomotor reflexes.*

Case 18. H.M. R.-V., F., 39 yr.

Refraction: R.E. AsH. $\begin{bmatrix} E \\ E - 2^1/_4 \, D. \end{bmatrix}$; L.E. HAsH. $\begin{bmatrix} E - 1/_2 \, D. \\ E - 2 \, D. \end{bmatrix}$

Vis. ac. with correction: R.E. $^1/_4$ with error; L.E. $^1/_2$ with error. Binoc. $^3/_4$ plus.

Very slight strab. conv. oc. dextr.; no alternating hyperphoria.

Latent nystagmus which had improved slightly in the course of years.

In binocular vision: straight ahead, no nystagmus; to R. and to L., terminal-position nystagmus; with a rotatory component on looking to the L.

In monocular vision with the R.E., jerking nystagmus to R.; amplitude greater on looking to R. and smaller on looking to L.

In monocular vision with the L.E., jerking nystagmus to L.; amplitude smaller on looking to R. and greater on looking to L.

No nystagmus in the dark.

Field of gaze normal.

Binocular perception and even some degree of depth perception were present. Binocular perception was achieved only slowly.

Very small fusion amplitude. Immediate suppression of one of the double images.

An abnormal binocular junction was demonstrated with Maddox' rods.

Optokinetic nystagmus

Binocular with movement to R.: jerking nystagmus to L.
   ,,   ,,   ,,   ,, L.:   ,,   ,,   ,, R.
Monocular R.E. with movement to R.: weak   jerking nystagmus to R.
   ,,   ,, ,,   ,,   ,,   ,, L.: strong   ,,   ,,   ,, R.
   ,,   L.E.   ,,   ,,   ,, R.:   ,,   ,,   ,,   ,, L.
   ,,   ,, ,,   ,,   ,,   ,, L.: weak   ,,   ,,   ,,   ,, L.

The phenomena observed in this case resemble those of the preceding case (Case 17) in many respects. In Case 18, however, there was no alternating hyperphoria. In both cases the nystagmus with monocular vision was influenced by optokinetic stimulation, but in neither case did a temporalwards movement of the contours succeed in completely overcoming the nasally-directed predominance of the gaze tonus. In Case 18 this was even more evident than in Case 17. Here thus, there was *a still rather more severe disturbance of the development of the conjugated optomotor reflexes.* The convergent strabismus, the labile binocular junction and the insufficient visual acuity indicate that *the development of the monocular optomotor reflexes was also disturbed.*

Although we do not propose to deal here with the optical localization, it must be pointed out that despite the presence of convergent strabismus with abnormal binocular junction (abnormal correspondence) the existence of depth perception could be demonstrated in this case.

Case 19. W.W., M., 28 yr.

Refraction: bilat. AsH. $\begin{bmatrix} E \\ E - 1\,D. \end{bmatrix}$ max. 5° temp.

Vis. ac. with correction: R.E. $^5/_4$; L.E. $1^1/_2 - {}^5/_4$; binoc. $1^1/_2$.

Strabismus convergens alternans; L.E. preferred; the strabismus had appeared after whooping cough at the age of 6 weeks.

Alternating hyperphoria.

Latent nystagmus with rotatory component.

No strabismus or nystagmus in the family.

When, with both eyes open, the R.E. fixated there was no difference in height; when the L.E. fixated the R.E. was higher.

In monocular vision with the R.E., weak jerking nystagmus to R.; this increased on looking to the R. and decreased on looking to the L.

In monocular vision with the L.E., rather stronger jerking nystagmus to L.; on looking to the R., rather irregular movements; on looking to the L. a stronger jerking nystagmus to the L.

No nystagmus in the dark.

The R.E. had some limitation of adduction, presumably as a result of a previous tenotomy; apart from this the field of gaze was normal.

No binocular perception; always alternating vision with immediate suppression of the image of the non-fixating eye.

Optokinetic nystagmus

Binocular with movement to R.: weak jerking nystagmus to L.
  „   „   „   „ L.:   „   „   „   „ R.
Monocular R.E. with movement to R.: weak   jerking nystagmus to R.
  „   „ „   „   „   „ L.: marked   „   „   „ R.
  „   L.E.   „   „   „ R.:   „   „   „   „ L.
  „   „ „   „   „   „ L.: weak   „   „   „ L.

On the grounds of the optokinetic nystagmus, which was identical with that in Case 18, we may conclude that here again there was *a rather more marked disturbance in the development of the conjugated optomotor reflexes.* But there were also certain differences; in Case 18 there was a terminal-position nystagmus but in Case 19 this was practically absent. Further, this patient had an *alternating hyperphoria,* which was absent in Case 18 but which was present in Case 17. Although the visual acuity was very good, both the strabismus and the alternating hyperphoria show that here also there must have been a *disturbance in the development of the monocular optomotor reflexes.*

Case 20. v. d. V.-T., F., 39 yr.

Refraction: HAsH. $\begin{bmatrix} E-8\,D. \\ E-9\,D. \end{bmatrix}$

Vis. ac. with correction: R.E. $^2/_5$; L.E. $^2/_5$.

Had formerly had strab. conv. oc. sin.; after operation the position of the eyes had been correct at first but strab. div. sursumverg. oc. dextr. had appeared later.

Alternating hyperphoria.

Latent nystagmus; had had pendular nystagmus in childhood.

No binocular perception; weakness of convergence, probably as a result of the operation.

Optokinetic nystagmus

Binocular with movement to R.: jerking nystagmus to L.

„   „   „   „ L.:   „   „   „ R.

Monocular R.E. with movement to R.: first no nystagmus; then jerking nystagmus to R.

Monocular R.E. with movement to L.: jerking nystagmus to R.

„   L.E.   „   „   „ R.:   „   „   „ L.

„   „ „   „   „   „ L.:   „   „   „ L.

The phenomena in this case, including the *alternating hyperphoria,* are again strongly reminiscent of those in Cases 17 and 19. Therefore we conclude once more that there was a *rather more severe disturbance in the development of the conjugated optomotor reflexes,* in addition to a *disturbance in the development of the monocular optomotor reflexes.* A point worthy of note in this case is that there had previously been a pendular nystagmus. Van der Hoeve (1917) also described the transition from a pendular nystagmus to a latent nystagmus. In Cases 2, 6 and 10 we noted that with monocular vision there was a somewhat stronger gaze tonus in the direction of the covered eye; these 3 patients, thus, had a suggestion of latent nystagmus in addition to their pendular nystagmus. It seems not improbable that in these patients too the pendular nystagmus would have disappeared if the fixation tonus from each eye in the direction of the other eye, i.e. the latent nystagmus, had developed further.

Case 21. C.D., M., 13 yr.

Refraction: bilat. hypermetropia E — 2 D.
Vis. ac. with correction: R.E. $^1/_{10}$ plus; L.E. $^5/_4$; binoc. $^5/_4$.
Strab. conv. oc. dextr.; no alternating hyperphoria.

Latent nystagmus; somewhat stronger when the R.E. was covered.

The boy tended to keep his head turned slightly to the left so that the eyes were directed a little to the R.

On looking to L. with both eyes open, coarse jerking nystagmus to L.; on looking to R. with both eyes open the eyes kept practically still.

No binocular depth perception.

Optokinetic nystagmus

Binocular with movement to R.: now and then a jerk to L.

,,　　,,　　　,,　　,, L.: eyes at rest.

Monocular R.E. with movement to R.: jerking nystagmus to R.

,,　　,, ,,　　,,　　　,,　　,, L.:　　,,　　　,,　　,, R.

,,　　L.E.　,,　　　,,　　,, R.:　　,,　　　,,　　,, L.

,,　　,, ,,　　,,　　　,,　　,, L.:　　,,　　　,,　　,, L.

Since the jerking nystagmus to the L. when the R.E. was covered was stronger than that to the R. when the L.E. was covered, the gaze tonus in this case cannot really be regarded as symmetrical. The asymmetry was also shown by the difference in terminal-position nystagmus on looking to the R. and to the L. and in the tendency to keep the head turned to the left. The gaze tonus to the L. appeared to be the weaker. However, as the optokinetic stimulation gave practically symmetrical results, we decided to place the case in this group. The predominant gaze tonus to the R. might be due to stimuli emanating from the R.E. or from the L.E. As the L.E. was the better eye and the optomotor reflexes from this eye had presumably developed better also, we feel that the predominance in question can be ascribed to a stronger fixation tonus to the R. from the L.E. From this it follows that a stronger jerking nystagmus with monocular vision does not necessarily mean that the optomotor reflexes from the seeing eye are less well developed. The jerking nystagmus is due to a *difference* in fixation tonus; if this fixation tonus is low in both directions or high in both directions, no strong nystagmus is to be expected. The simple fact of a stronger jerking nystagmus in monocular vision does not in itself prove that the development of the optomotor reflexes from that eye was better. Nevertheless, the fixation tonus in this case was in general very low. This is shown by the optokinetic nystagmus. Up to now we had noted in all our cases of latent nystagmus a normal optokinetic nystagmus with binocular vision; this patient, however, showed little or no reaction to optokinetic stimulation ('optische Drehstarre').

In this case, thus, we are concerned with a *severe disturbance in the development of the conjugated optomotor reflexes.* The strabismus shows that there was also *a disturbance in the development of the monocular optomotor reflexes.*

Case 22. W.J., M., 17 yr.

Refraction: R.E. HAsH. $\begin{bmatrix} E - 1^1/_2 \, D. \\ E - 3 \, D. \end{bmatrix}$ max. 10° temp.

L.E. HAsH. $\begin{bmatrix} E - 1^1/_2 \, D. \\ E - 3^1/_2 \, D. \end{bmatrix}$ max. 10° temp.

Vis. ac. with correction R.E. $^3/_5$; L.E. $^3/_5$.

Formerly strab. conv. oc. dex.; later strab. conv. oc. sin.; eyes now parallel with spectacles. Strabismus had appeared before the age of 6 months. The head was kept turned slightly to the R., thus the eyes to the L.

No alternating hyperphoria.

Latent nystagmus.

Strabismus and nystagmus in the family.

On looking to the R. or L. with both eyes, coarse terminal-position nystagmus.

In monocular vision with the R.E.: straight ahead, jerking nystagmus to R.; to the R., strong jerking nystagmus to R.; to the L., very weak jerking nystagmus to R. or pendular nystagmus.

In monocular vision with the L.E.: straight ahead, jerking nystagmus to L.; to the R., very weak jerking nystagmus to R.; to the L., strong jerking nystagmus to L.

In the dark, slow to and fro wavering of the eyes.

Field of gaze normal.

Good binocular perception, also depth perception; incorrect estimation of size.

Small fusion amplitude; direct suppression of one of the double images.

Optokinetic nystagmus

Binocular with movement to R.: eyes still at first; then jerking nystagmus to R.

Binocular with movement to L.: weak jerking nystagmus to R.

Monocular R.E. with movement to R.: weak   jerking nystagmus to R.

   „   „ „   „   „   „ L.: strong   „   „   „ R.

   „   L.E.   „   „   „ R.:   „   „   „ L.

   „   „ „   „   „   „ L.:   „   „   „ L.

As in Case 21, we see here a slight asymmetry in the gaze tonus; there is a slight predominance of the fixation tonus to the L. This is also shown by the position of the head, the terminal-position nystagmus with use of the R.E. and the opto-kinetic nystagmus with binocular vision. The difference in fixation tonus was greatest in monocular vision with the R.E.; which was also the preferred eye. Probably the leftward con-jugated optomotor reflexes from this R.E. had developed some-what better. Further, the fixation tonus was in this case also very low. With binocular vision and movement of the objects to the R. we even obtained an inverse type. This case is as it were a mirror-image of Case 21, while being further identical with the latter in practically all respects. We may conclude that here again there was a *severe disturbance in the development of the conjugated optomotor reflexes* and a *less severe disturbance in the development of the monocular optomotor reflexes.*

Case 23. E. d. B., M., 8 yr.

Refraction: R.E. MAsM. $\begin{bmatrix} E + 6^1/_2 \text{ D.} \\ E + 4^1/_2 \text{ D.} \end{bmatrix}$ max. vertic.;

L.E. MAsM. $\begin{bmatrix} E + 6^1/_2 \text{ D.} \\ E + 4^1/_2 \text{ D.} \end{bmatrix}$ max. $10°$ temp.

Vis. ac. with correction: R.E. $^4/_5$; L.E. $^4/_5$.

Strabismus convergens alternans periodicus. No alternating hyperphoria.

The strabismus had almost disappeared in the course of years.

Latent nystagmus.

No strabismus or nystagmus in the family.

With binocular vision: straight ahead, quivering of the eyes, sometimes with a few jerks; to the R. and to the L. marked terminal-position nystagmus.

In monocular vision with the R.E.: straight ahead, jerking nystagmus to R.; to the R., strong jerking nystagmus to R.; to the L., very weak jerking nystagmus to L.

In monocular vision with the L.E.: straight ahead, fine jerking nystagmus to L.; to the R., jerking nystagmus to R.; to the L., coarse jerking nystagmus to L.

In the dark, fine jerking nystagmus to R. and L. alternately.

Field of gaze normal.

There was a suggestion of binocular single vision (simul-taneous perception); very rapid suppression of the left retinal image.

Optokinetic nystagmus

Binocular with movement to R.: quivering of the eyes; practically at rest.

„ „ „ „ L.: „ with a few jerks to R.

Monocular R.E. with movement to R.: sometimes eyes still, sometimes jerks to R.

Monocular R.E. with movement to L.: jerks to R.

„ L.E. „ „ „ R.: eyes still or jerks to L.

„ „ „ „ „ „ L.: jerking nystagmus to L.

The nystagmus in the dark shows that the non-optical gaze tonus was insufficient to keep the eyes still. Since the optomotor reflexes were very poorly developed, it is probable that the non-optical gaze tonus had also suffered as a result of this (see also Cases 4 and 6). The very poor development of the optical fixation tonus is shown by the quivering of the eyes in binocular vision, the strong terminal-position nystagmus and more especially the reaction to optokinetic stimulation with monocular vision. In monocular vision with the R.E. and movement of the objects to the L. the jerking nystagmus to the R. did not increase but decreased to weak jerks to the R., i.e. an inverse type. This was still clearer in monocular vision with the L.E.; movement of the objects to the R. did not cause the existing jerking nystagmus to L. to increase but brought the eyes practically to a standstill. Movement of the objects to the L., on the other hand, rather accentuated than reduced the jerking nystagmus to the L. On the grounds of this *inverse type* we can certainly say that there was a *severe disturbance of the development of the conjugated optomotor reflexes.* The tendency to strabismus also points, despite the fairly good visual acuity, to *some degree of disturbance in the development of the monocular optomotor reflexes.*

Case 24. A. d. V., M., 39 yr.

Refraction: R.E. M. E $+ 2^1/_2$ D.; L.E. MAsM. $\left[\begin{array}{l} \text{E} + 3^1/_2 \text{ D.} \\ \text{E} + 2^1/_2 \text{ D.} \end{array} \right.$

Vis. ac. with correction: R.E. $^5/_4$; L.E. $^3/_4$; binoc. $^5/_4$.

No strabismus; no alternating hyperphoria.

Latent nystagmus. Nystagmus in the family.

With binocular vision: straight ahead no nystagmus; to the R. and to the L., terminal-position nystagmus.

In monocular vision with the R.E.: straight ahead, jerking

nystagmus to R.; to the R., increasing jerking nystagmus to R.; to the L., disappearance of the jerking nystagmus.

In monocular vision with the L.E.: straight ahead, jerking nystagmus to L.; to the R., decrease of the jerking nystagmus to L.; to the L., increase of the jerking nystagmus to L.

Simultaneous perception but no depth perception.

Very small fusion amplitude with prompt suppression of one of the double images.

Optokinetic nystagmus

Binocular with movement to R.: pendular nystagmus.

"   "   "   " L.:   "   "

Monocular R.E. with movement to R.: jerking nystagmus to R.

"   " "   "   "   " L.:   "   "   " R.

"   L.E.   "   "   " R.:   "   "   " L.

"   " "   "   "   " L.:   "   "   " L.

The optokinetic stimulation with binocular vision weakened the already insufficient fixation tonus that was present without such stimulation. This exhaustion is only understandable if the fixation tonus was very low indeed. With monocular vision there was no sign whatever of any influence of the optokinetic stimulation. Since the predominating fixation tonus which gave rise to the jerking nystagmus with monocular vision can only have been low, as shown by the optokinetic pendular nystagmus with binocular vision, the fixation tonus in the direction of the open eye with monocular vision must have been practically nil. This also accounts for the 'optische Drehstarre' with monocular vision.

Taking all these points into account we must conclude that this patient had a *severe disturbance in the development of the conjugated optomotor reflexes.* The poor binocular perception and the very small fusion amplitude also suggest a *slight disturbance in the development of the monocular optomotor reflexes.*

Case 25. G.S., M., 22 yr.

Refraction: bilateral hypermetropia E — 1 D.
Vis. ac. with correction: R.E. $^1/_{10}$ plus; L.E. $^3/_5$.
Strabismus divergens oc. dex. No alternating hyperphoria.
Bilateral posterior subcapsular cataract.

With binocular vision: straight ahead, now and then pendular nystagmus with jerks to R.; to the R. and to the L., terminal-position nystagmus. In monocular vision with the R.E., marked jerking nystagmus to R., becoming still stronger on looking to the R. and decreasing on looking to the L.

In monocular vision with the L.E., marked jerking nystagmus to the L., decreasing on looking to the R. and increasing on looking to the L.

No binocular perception; a trace at the most of simultaneous perception.

Optokinetic nystagmus

Binocular with movement to R.: jerking nystagmus to R.

     „   „   „   „ L.:   „   „   „ L.

Monocular R.E. with movement to R.: jerking nystagmus to R.

   „   „ „   „   „   „ L.:   „   „   „ R.

   „   L.E.   „   „   „ R.:   „   „   „ L.

   „   „ „   „   „   „ L.:   „   „   „ L.

The inverse type with optokinetic stimulation when both eyes were open indicates a pronounced deficiency of the fixation tonus. This cannot, however, have been completely absent, as in that case it would be impossible to account for the strong jerking nystagmus in monocular vision. From these considerations it follows that in monocular vision the fixation tonus in the direction of the open eye must have been practically nil, as in the previous patient (Case 24). Therefore in monocular vision the jerking nystagmus could not be inhibited by temporalwards movement of the contours. Here again there was a *severe disturbance in the development of the conjugated optomotor reflexes* and, as indicated by the strabismus and amblyopia, also a *disturbance in the development of the monocular optomotor reflexes*.

Case 26. J.K., F., 8 yr.

Refraction: bilateral hypermetropia E — 5 D.

Vis. ac. with correction: R.E. 1; L.E. 1; binoc. 1.

Strab. conv. oc. sin. from before the age of 6 months, subsequently changed to strabismus convergens alternans. At the time of examination the position of the eyes was parallel with spectacles. No alternating hyperphoria.

With binocular vision in daylight, very fine pendular nystagmus; further nystagmus latens. There had formerly been a well-marked pendular nystagmus.

With binocular vision: to the R., fine jerking nystagmus to R.; to the L., coarse jerking nystagmus to L.

In monocular vision with the R.E., weak jerking nystagmus to R., increasing on looking to the R. and disappearing on looking to the L.

In monocular vision with the L.E., somewhat stronger jerking nystagmus to L., increasing on looking to the L. and decreasing on looking to the R.

Binocular perception with a suggestion of depth perception.

Optokinetic nystagmus

Binocular with movement to R.: very fine pendular nystagmus.
    „   „   „   „ L.: a few small jerks to R.
Monocular R.E. with movement to R.: coarse jerking nystagmus to R.
    „   „ „   „   „ L.: eyes practically at rest.
    „   L.E.   „   „   „ R.: „   „   „   „
    „   „ „   „   „ L.: coarse jerking nystagmus to L.

As in Cases 21 and 22 there was here a slight asymmetry of the optical gaze tonus. There was a slight preponderance of the fixation tonus to the R., as a result of which the terminal-position nystagmus on looking to the L. was somewhat stronger, while the jerking nystagmus in monocular vision with the L.E. was also stronger than that in monocular vision with the R.E. In this patient, whose fixation tonus was further very low, the predominance must have come from stimuli from the L.E. The previous pendular nystagmus shows that at first the fixation tonus must have been practically nil (see also Case 20). A residue of this pendular nystagmus had remained in the form of a fine quivering of the eyes (see also Cases 23 and 25). The jerking nystagmus in monocular vision shows that a fixation tonus must have gradually developed: from the R.E. a fixation tonus to the L. and from the L.E. a somewhat stronger fixation tonus to the R. In monocular vision a temporalwards movement of the contours had no inhibitory influence on the jerking nystagmus but rather tended to accentuate it (typus inversus), probably owing to exhaustion of a minimal temporally-directed fixation tonus. With nasalwards movement of the contours the existing jerking nystagmus disappeared; this typus inversus also can only be accounted for by exhaustion, but now of the nasally directed fixation tonus (see also Case 23). The fixation tonus must have been very low in both directions, both temporal and nasal. But what had caused the almost complete disappearance of the pendular nystagmus in binocular vision? This had in fact been present for several years; probably its disappearance was partly due to the *fairly good although delayed development of the monocular optomotor reflexes,* as demonstrated by the good visual acuity and the binocular perception. Against this we have the fact that the inverse type of the optokinetic nystagmus proves the existence of a *very severe disturbance of the development of the conjugated optomotor reflexes.*

Case 27. G.W., M., 7 yr.

Refraction: bilat, hypermetropia E — 3 D.

Vis. ac. with correction: R.E. 1; L.E. 1; binoc. 1.

Strabismus convergens alternans; L.E. preferred. Strabismus present at the age of 7 months. Formerly strabismus convergens sursumvergens. At the time of examination the sursumvergence had disappeared; thus there had formerly been an alternating hyperphoria.

Congenital pendular nystagmus, increasing with monocular vision. At the time of examination only latent nystagmus remained. Here again we see a pendular nystagmus that had changed into a latent nystagmus. Nothing was known as to familial occurrence.

With binocular vision: straight ahead, eyes at rest; to the R. and to the L., coarse terminal-position nystagmus.

In monocular vision with the R.E.: straight ahead, jerking nystagmus to R.; to the R., coarse jerking nystagmus to R.; to the L. eyes at rest.

In monocular vision with the L.E.: straight ahead, jerking nystagmus to L.; to the R., eyes at rest; to the L., coarse jerking nystagmus to L.

In the dark, pendular nystagmus.

Field of gaze normal.

Reasonably good binocular perception with depth perception; judgement of sizes bad.

Fusion amplitude approx. 8°; very rapid suppression of one of the retinal images.

Optokinetic nystagmus

Binocular vision with movement to R.: jerking nystagmus to R.

    ,,   ,,   ,,   ,,   ,, L.:   ,,   ,,   ,, L.

Monocular R.E. with movement to R.: very weak jerking nystagmus to L.

    ,,   ,, ,,   ,,   ,,   ,, L.: jerking nystagmus to L.

    ,,   L.E.   ,,   ,,   ,, R.: definite jerking nystagmus to R.

    ,,   ,, ,,   ,,   ..   ,, L.: only a few jerks to R.

Nystagmus in the dark was observed also in Cases 2, 3, 4, 6, 11, 13, 22 and 23. In such cases the non-optical reflexes are not capable of keeping the eyes still; there is an insufficiency not only of the optical but also of the non-optical gaze tonus. The question is whether the disturbance of the non optical reflexes or that of the optical reflexes is primary, or whether both are the result of a more generalized development disturbance. We cannot supply a satisfactory answer to this question. It must be remembered, however, that persons born blind also show a kind of nystagmus even though there may be no evidence of a disturbance in the non-optical reflexes. Normal individuals with good eyes have no nystagmus in the dark. From these considerations we feel justified in concluding, as we have also explained in the introduction, that optical stimuli may enhance the non-optical gaze tonus. If the optical stimuli are entirely lacking, the non-optical gaze tonus will also remain deficient.

The optokinetic nystagmus in this case showed an unmistakable inverse type both in binocular and in monocular vision. Normally one would expect that the jerking nystagmus in monocular vision would be accentuated by a nasalwards movement of the contours, but what actually appeared was a jerking nystagmus in the opposite direction. However we may try to explain the inverse type, by exhaustion or in other ways, it will only appear if the optical fixation reflexes in the direction of the moving objects are very poorly developed. This accounts for both the nystagmus in the dark and the inverse type. What it does not account for, however, is the appearance of nasalwards jerks with monocular vision when the contours moved in the temporal direction. A normal production of optokinetic nystagmus by an increased fixation tonus in the temporalward direction cannot be expected. We are more inclined to suggest an inhibition of the small residue of the nasally directed optical tonus, as a result of which the eyes — perhaps partly owing to a persisting tonus from de covered eye — slip somewhat in a temporal direction and endeavour by means of small adjusting movements to maintain the original position.

In any case, we were here concerned with a *very severe disturbance of the development of the conjugated optomotor reflexes,* while the *monocular optomotor reflexes were less defective* (visual acuity; depth perception).

In daylight the optical reflexes were still just capable of suppressing the tendency to nystagmus with the patient looking straight ahead.

Case 28. H.J., M., 26 yr.

Refraction: R.E. AsM. $\begin{bmatrix} E \\ E + 2\,D. \end{bmatrix}$;

L.E. As. mixt. $\begin{bmatrix} E - 3\,D. \\ E + 1\,D. \end{bmatrix}$ max. 70° temp.

Vis. ac. with correction: R.E. $^2/_5$; L.E. $^2/_5$.

No strabismus; no alternating hyperphoria.

Pendular nystagmus and latent nystagmus. Formerly jerking nystagmus to L. with rotatory component.

With binocular vision: straight ahead, pendular nystagmus; to the R., jerking nystagmus to R.; to the L., weak jerking nystagmus to L.

In monocular vision with the R.E.: straight ahead, slight jerking nystagmus to R.; to the R., marked jerking nystagmus to R.; to the L., fine jerking nystagmus to L.

In monocular vision with the L.E.: straight ahead, weak jerking nystagmus to L.; to the R., pendular nystagmus; to the L., strong jerking nystagmus to L.

In the dark, pendular nystagmus.

Field of gaze normal.

Good binocular perception; also depth perception; estimation of sizes normal.

Normal fusion amplitude; no rapid suppression.

Optokinetic nystagmus

Binocular with movement to R.: jerking nystagmus to R.

   ,,   ,,   ,,   ,, L.: ,,   ,,   ,, L.

Monocular R.E. with movement to R.: pendular nystagmus.

   ,,   ,, ,,   ,,   ,,   ,, L.:   ,,   ,,

   ,,   L.E.   ,,   ,,   ,, R.:   ,,   ,,

   ,,   ,, ,,   ,,   ,,   ,, L.:   ,,   ,,

The phenomena shown by this patient were in many respects similar to those of the previous one (Case 27). In this case however, there was also a pendular nystagmus in daylight with binocular vision, showing that the residue of the optical conjugated reflexes was so small as to be incapable of suppressing the pendular nystagmus. The inverse type with optokinetic stimulation and binocular vision is in perfect agreement with this. In monocular vision, however, the optokinetic stimulation invariably elicited a pendular nystagmus, whatever the direction of movement of the objects. A pendular nystagmus means that the optical fixation reflexes either just balance each other, although they are insufficient, or are completely absent. The latter seems to be the case here. If, as a result of optokinetic stimulation with monocular vision, the minimal fixation tonus in the direction of the covered eye is exhausted or inhibited, nothing at all remains. On the grounds of these phenomena we conclude that this patient had an *extremely severe form of developmental disturbance in the conjugated optomotor reflexes.*

The poor visual acuity suggests also a *disturbance in the development of the monocular optomotor reflexes,* at any rate those from the central portions of the retinae; this disturbance, however, cannot have been severe, in view of the very good binocular perception, normal fusion amplitude and absence of strabismus.

Summary of Group III.

Fifteen cases with latent nystagmus and practically symmetrical optical gaze tonus are described. In all these cases there was, of course, a disturbance of the optical fixation tonus. The degree of such disturbance could be satisfactorily judged from the greater or less degree of terminal-position nystagmus, or still better from the reaction to optokinetic stimulation with binocular and with monocular vision. In Cases 14, 15, 16, 17, 19 and 20 there was no appreciable terminal-position nystagmus, whereas Cases 18, 21, 22, 23, 24, 25, 26, 27 and 28 did show this phenomenon. Turning now to the optokinetic nystagmus with binocular vision we find a normal, though sometimes rather weak optokinetic nystagmus in Cases 14, 15, 16, 17, 18, 19 and 20. No reaction to optokinetic stimulation with binocular vision was seen in Cases 21, 22, 23 and 26. An inverse type with binocular vision was seen in Cases 25, 27 and 28 while Case 24 responded with a pendular nystagmus. With monocular vision the findings were as follows: Normal optokinetic nystagmus was seen only in Case 15. A weakening of the existing jerking nystagmus with temporalwards movement of the contours and an accentuation thereof with nasal movement of the contours was shown by Cases 16, 17, 18, 19, 20 and

22. A practically complete absence of any influence of opto-kinetic stimulation was seen in Cases 14, 21, 24 and 25, an inverse type in Cases 23, 26 and 27 and a pendular nystagmus in Case 28.

On the basis of the above observations we were able to arrange our cases very satisfactorily in order of severity of their anomalies. The nystagmus in the dark could also have served as a criterion, but this was not investigated in all cases; however, with the exception of Case 24, Cases 22 to 28 inclusive all showed nystagmus in the dark.

A phenomenon which was frequently associated with latent nystagmus was strabismus. Only Cases 14, 24 and 28 did not squint and, as far as could be ascertained, had never done so. Five patients had an alternating hyperphoria or a suggestion thereof (Cases 15, 17, 19, 20 and 27).

Only 6 patients had satisfactory binocular perception (Cases 14, 15, 18, 26, 27 and 28). This shows the notable feature that binocular perception is not necessarily altered with the disturbance in the optical fixation tonus. Cases 23 and 24 had no depth perception but did have some degree of simultaneous perception. A notable feature, also in the non-squinting patients, was the small fusion amplitude and the great tendency to suppression of one of the retinal images. Only in Case 28 — who had the lowest fixation tonus of all — was there a normal binocular perception and normal fusion amplitude. Thus the fusion movement is, in our opinion, not dependent on the conjugated optomotor reflexes but more intimately connected with the monocular optomotor reflexes and the cortical binocular junction.

Cases 20, 26 and 27 had formerly had a pendular nystagmus, which had disappeared as the latent nystagmus developed; in Case 26, however, a trace of the pendular nystagmus still remained. In the development of latent nystagmus there is established a fixation tonus from each eye in the direction of the other eye; this makes it possible for the eyes to be kept still as long as they are both open. If this establishment of fixation tonus is not equal in both eyes, we get larger or smaller deviations from the symmetrical gaze tonus, as seen in Cases 21, 22 and 26.

On the grounds of our findings we believe the latent nystagmus to be caused by a defective fixation tonus; this defective fixation tonus we ascribe to a disturbance in the development of the conjugated optomotor reflexes. In many cases we also found a disturbance in the development of the monocular optomotor reflexes; but if we take strabismus, visual acuity, fusion amplitude and binocular perception into account, we find that the latter disturbance does not by any means always run parallel to the disturbance in the conjugated reflexes.

Group IV. Patients with latent nystagmus and asymmetric optical gaze tonus.

This group comprises a number of patients with latent nystagmus whose optical gaze tonus was more or less asymmetric. In the preceding group we also included 3 patients (Cases 21, 22 and 26) who had a very slightly asymmetric gaze tonus. This asymmetry means that the gaze tonus elicited from the R.E. differs from that elicited from the L.E. This may be due to a better development of the nasally directed optical fixation tonus in one eye, but it can also be due to the fact that in one eye both the temporally and the nasally directed optical fixation tonus have developed better and more equally.

Case 29. G.d.L., F., 16 yr.

Refraction: R.E. AsH. $\left[\begin{array}{l} E - 1\,D. \\ E \end{array}\right.$ max. $60°$ temp.;

L.E. H. E — 1 D.

Vis. ac. with correction: R.E. $^3/_4$ plus; L.E. $^1/_{10}$ with error; binoc $^3/_4$ plus.

Strabismus convergens sursumvergens oc. sin.; head inclined somewhat to R. shoulder.

Suggestion of alternating hyperphoria.

Strabismus noticed at the age of 3 weeks.

Rotatory pendular nystagmus and latent nystagmus.

No strabismus or nystagmus in the family.

With binocular vision: straight ahead irregular rotatory movements; to the R., jerking nystagmus to R. and upwards (the latter especially with the L.E.); to the L., rotatory pendular nystagmus.

In monocular vision with the R.E.: straight ahead, slight jerking nystagmus to R. and slightly upwards; to the R., jerking nystagmus to R.; to the L., rotatory pendular nystagmus.

In monocular vision with the L.E.: rotatory pendular nystagmus alternating with jerking nystagmus to L.; on looking to the R. no nystagmus; to the L., rotatory pendular nystagmus with jerks to L.

Some limitation of abduction of the L.E.; field of gaze further normal.

No binocular perception or simultaneous perception.

Optokinetic nystagmus

Binocular with movement to R.: jerking nystagmus to L.

    „   „   „   „ L.: R.E. jerking nyst. to R.; L.E. turned upward.

Monocular R.E. with movement to R.: no nystagmus.

    „   „ „   „   „   „ L.: good jerking nystagmus to R.

    „   L.E.   „   „   „ R.: weak   „   „   „ L.

    „   „ „   „   „   „ L.: no nystagmus.

The terminal-position nystagmus, which was stronger on looking to the R., showed that there was some predominance of the fixation tonus to the L. This predominance was obviously due to stimuli from the R.E.; in monocular vision with the L.E. the jerking nystagmus was only periodically present, while on looking to the L. the rotatory pendular movement still preponderated.

The fixation tonus from the better R.E. was more strongly developed than that from the worse L.E.: as a result of this the difference in fixation tonus for R. and L. turning in monocular vision with the R.E. was also more pronounced and the jerking nystagmus stronger. Optokinetic stimulation with monocular vision was not without influence. With temporalwards movement the existing jerking nystagmus was abolished; with nasalwards movement it was accentuated, at any rate for the R.E. with its higher fixation tonus.

Case 30: J.W.S., F., 20 yr.

Refraction: R.E. H.AsH. $\left[\begin{array}{l} E - 1^{1}/_{2} \text{ D.} \\ E - 3^{3}/_{4} \text{ D.} \end{array}\right]$;

$\qquad\qquad\qquad$ L.E. HAsH. $\left[\begin{array}{l} E - 1^{3}/_{4} \text{ D.} \\ E - 4 \text{ D.} \end{array}\right]$

Vis. ac. with correction: R.E. $^{3}/_{4}$ plus; L.E. $^{1}/_{3}$ with errors; binoc. $^{3}/_{4}$ plus.

Strabismus convergens oc. sin.; suggestion of strab. sursoadductorius.

Very slight alternating hyperphoria.

Strabismus first noticed at the age of about 7 months.

Latent nystagmus.

No strabismus or nystagmus in the family.

With binocular vision: straight ahead, sometimes fine jerks to R.; to the R., jerks to R.; to the L., eyes sometimes still, sometimes jerks to R. or L.

In monocular vision with the R.E.: straight ahead, jerking nystagmus to R.; to the R., jerking nystagmus increased; to the L., jerking nystagmus greatly decreased.

In monocular vision with the L.E.; straight ahead, jerking nystagmus to L.; to the R., jerking nystagmus to R.; to the L., strong jerking nystagmus to L.

Field of gaze normal.

Simultaneous perception achieved now and then; no depth perception.

Very prompt suppression of one of the retinal images.

Abnormal cortical binocular junction (abnormal correspondence).

Optokinetic nystagmus

Binocular with movement to R.: very weak jerking nystagmus to L.

    ,,   ,,   ,,   ,, L.: jerking nystagmus to R.

Monocular R.E. with movement to R.: pendular nystagmus; a few jerks to L.

Monocular R.E. with movement to L.: well-marked jerking nystagmus to R.

Monocular L.E. with movement to R.: good jerking nystagmus to L.

    ,,   ,, ,,   ,,   ,,   ,, L.: weak   ,,   ,,   ,, L.

The fine jerks to R. with both eyes open indicate a predominance of the fixation tonus to L. This was also in evidence on looking to the R. or L., as only looking to the R. gave a true terminal-position nystagmus. The difference in fixation tonus for R. or L. turning was greater in monocular vision with the R.E. than in monocular vision with the L.E. When the R.E. moved to the R. or L. in monocular vision, the jerking nystagmus to R. was preserved. When the L.E. moved to the R. or L. in monocular vision, the jerking nystagmus changed its direction.

The difference in fixation tonus is also shown by the optokinetic nystagmus. With binocular vision and movement of the contours to the R. there was only a weak jerking nystagmus to L., because the predominance of the tonus to L. had first to be overcome. The influence of optokinetic stimulation was very marked in monocular vision with the R.E. and much weaker in monocular vision with the L.E. Here again we see, as in Case 29, that the *fixation tonus from the better R.E., especially in the nasal direction, had developed better than the fixation tonus from the poorer L.E.*

Case 31. Th.E., M., 41 yr.

Refraction: R.E. Hm. E — 16 D.; L.E. Hm. E — 14 D.
Vis. ac. with correction: R.E. $^1/_{300}$; L.E. $^5/_{60}$; binoc. $^5/_{60}$.
Strab. conv. oc. dex. from earliest childhood.
No alternating hyperphoria.
R.E.: microphthalmus, aphakia, maculae corneae, atrophia iridis, opacities in the vitreous; fundus atrophy as in myopia.
L.E.: VIth nerve paresis, aphakia, glaucoma simplex; clear vitreous; papilla slightly excavated with atrophic zone; field of vision intact.
Latent nystagmus.
With binocular vision: straight ahead, no nystagmus as a rule, sometimes a few jerks to L.; to the R., no nystagmus; to the L., jerking nystagmus to L.
In monocular vision with the R.E.: alternately eyes still and slight jerking nystagmus to R.; on looking to the R., jerking nystagmus to R.; to the L., no nystagmus.
In monocular vision with the L.E.: slight jerking nystagmus to L.; on looking to the R., very slight jerking nystagmus to L.; to the L., somewhat stronger jerking nystagmus to L.
No nystagmus in the dark.
Field of gaze of L.E. limited temporally.
No binocular perception.

Optokinetic nystagmus

Binocular with movement to R.: jerking nystagmus to L.
    ,,   ,,   ,,   ,, L.:   ,,   ,,   ,, R.

Monocular R.E. with movement to R.: ⎤ eyes still or slight jerking nystag-
   ,,   ,, ,,   ,,   ,, L.: ⎦ mus to R.; movement of contours barely perceived.

   ,,   L.E.   ,,   ,,   ,, R.: strong jerking nystagmus to L.
   ,,   ,, ,,   ,,   ,, L.: eyes still.

The jerks to L. with binocular vision indicate a predominance of the conjugated optical fixation tonus to the R. This predominance was due to stimuli from the less bad L.E. This was shown in monocular vision. In monocular vision with the R.E. the jerking nystagmus was weaker than with monocular vision with the L.E. On looking to the R. or L. with the R.E. the jerking nystagmus was in the direction of gaze. On looking to the R. or L. with the L.E. the jerking nystagmus was maintained. All this points to a predominance of the fixation tonus to R. from the L.E. Nevertheless, in monocular vision with the R.E., despite its bad visual acuity, there was still a predominance of the fixation tonus to L. Although this was small, it had the consequence that optokinetic stimulation with both eyes open gave a reaction different from that with the R.E. covered. When the R.E. was not occluded the predominance of fixation tonus from the L.E. to R. was partially compensated from the R.E. and we got a fairly normal optokinetic nystagmus. When the R.E. was occluded this compensation was lacking and the contour-movement to the L. had first to overcome the predominance of tonus to the R., so that the eyes came to rest; whereas a contour-movement to R. gave an unmistakable accentuation of the existing jerking nystagmus. This supports the view that the optomotor reflexes responsible for the fixation tonus are probably not the same as those which form the basis of visual acuity. As in the 2 preceding cases, we may conclude here that the *fixation tonus, and more especially the fixation tonus to the R., from the better L.E. had developed further than the fixation tonus from the very bad R.E.*

Case 32. J. Sch., M., 25 yr.

Refraction: Emmetr. Vis. ac.: R.E. $^5/_4$; L.E. $^1/_{10}$; binoc. $^5/_4$. Strab. conv. oc. sin. from the age of 1 yr.
No alternating hyperphoria.
Bilateral cataracta zonularis.
Latent nystagmus.
In binocular vision to the R. or L., terminal-position nystagmus.
In monocular vision with the R.E., jerking nystagmus to R. with a rotatory component.
In monocular vision with the L.E., only an occasional jerk to L.
No binocular perception; partial simultaneous perception but no fusion of the double images; prompt suppression of one of the double images.

Optokinetic nystagmus

Binocular with movement to R.: eyes at rest.
„   „   „   „ L.:   „   „   „
Monocular R.E. with movement to R.: fine jerking nystagmus to R.
„   „ „   „   „   „ L.: jerking nystagmus to R.
„   L.E.   „   „   „ R.:   „   „   „ L.
„   „ „   „   „   „ L.: eyes at rest.

The obvious conclusion would seem to be that this patient's cataracta zonularis had interfered with the normal development of the optomotor reflexes. But the possibility must also be considered that one and the same harmful agency might have caused both the cataracta zonularis and the disturbance in development of the optomotor reflexes. There was, in any case, a rather severe disturbance of the development of the conjugated optomotor reflexes, as shown by the fact that with binocular vision there was little or no reaction to optokinetic stimulation ('optische Drehstarre'). The latent nystagmus and the optokinetic nystagmus with monocular vision showed that in monocular vision there was a predominance of the nasally directed gaze tonus. This predominance was very marked in monocular vision with the good R.E. and extremely slight in monocular vision with the amblyopic L.E. The fixation tonus to R. from the L.E. was also poorly developed. Here again we conclude that the *fixation tonus from the better R.E., especially in the direction of the other eye, had developed better than that from the worse L.E.*

Case 33. N. v. d. B., M., 18 yr.

Refraction: myopia levior; E $+ \, ^1/_2$ D.
Vis. ac. with correction: R.E. $^1/_3$ with error; L.E. $^1/_{10}$; binoc. $^3/_4$ with error.

No strabismus; no alternating hyperphoria.

Sluggishly reacting pupils. Papillae rather pale, especially L.

Fields of vision somewhat reduced nasally. Had had severe convulsions shortly after birth.

Latent nystagmus.

With binocular vision: straight ahead, no nystagmus or fine jerking nystagmus to R.; to the R., jerking nystagmus to R.; to the L., no nystagmus or fine jerking nystagmus to L.

In monocular vision with the R.E.: straight ahead, jerking nystagmus to R.; to the R., jerking nystagmus to R.; to the L., no nystagmus or jerks to R.

In monocular vision with the L.E.: straight ahead, jerking nystagmus to L.; to the R., weak jerking nystagmus to L.; to the L., strong jerking nystagmus to L.

In the dark, slight jerking nystagmus to R., gradually disappearing.

Field of gaze normal.

Simultaneous perception but no depth perception.

Very small fusion amplitude, difficult to measure because with unlike images the left retinal image was immediately suppressed.

Convergence very defective; L.E. deviated to temporal almost at once.

Optokinetic nystagmus

Binocular with movement to R.: no nystagmus.
,, ,, ,, ,, L.: jerking nystagmus to R.
Monocular R.E. with movement to R.: jerking nystagmus to R.
,, ,, ,, ,, ,, ,, L.: ,, ,, ,, R.
,, L.E. ,, ,, ,, R.: ,, ,, ,, L.
,, ,, ,, ,, ,, ,, L.: no nystagmus or slight jerking nystagmus to L.; sometimes also pendular nystagmus.

Predominance of the fixation tonus to L. was demonstrated by the fine jerks to R. in binocular vision and by the optokinetic nystagmus with binocular and monocular vision. In monocular vision with the R.E., the predominance of the fixation tonus to L. could not be overcome by movement of the contours to R. In monocular vision with the L.E., however, the predominance of the fixation tonus to R. could be abolished by contour movement to L. From this it also follows that the predominance to the L. was caused by stimuli from the R.E. Since the L.E. was amblyopic, we can safely assume that the optomotor reflexes from the R.E. had developed better, so that the asymmetry must be ascribed to a *better development of the fixation tonus, especially in the direction of the other eye, from the better R.E. and a poorer development of the fixation tonus from the worse L.E.*

Case 34. C.H., F., 4 yr.

Refraction: bilat. hypermetropia E — 1 D.

Vis. ac. with correction: R.E. ?; L.E. 1; binoc. 1.

Strab. conv. oc. dex.; sometimes alternating; practically from birth.

Convergence highly variable; the R.E. often slightly higher. No alternating hyperphoria.

Cataracta congenita centralis oc. dextr.

Slight latent nystagmus; formerly pendular nystagmus.

With binocular vision to R., very slight jerking nystagmus of the L.E. to R. while the R.E. executed irregular movements; with binocular vision to L., coarse jerking nystagmus to L.

In monocular vision with the R.E., very irregular movements. In monocular vision with the L.E., weak jerking nystagmus to L.

In the dark also slight irregular nystagmus.

Optokinetic nystagmus

Binocular with movement to R.: jerking nystagmus to L.

     „   „   „   „ L.: irregular wavering movements.

Monocular R.E. with movement to R.: weak jerking nystagmus to R.

     „   „ „   „   „   „ L.: irregular wavering movements.

     „   L.E.   „   „   „ R.: appreciable jerking nystagmus to L.

     „   „ „   „   „   „ L.: irregular wavering.

In this case we have predominance of the fixation tonus to the R., as shown by the terminal-position nystagmus and by the optokinetic nystagmus with binocular vision. This predominance was due to stimuli from the L.E., as can be deduced from the jerking nystagmus in monocular vision with the L.E., which was absent in monocular vision with the R.E. This is also shown by the optokinetic nystagmus with monocular vision. The R.E. showed either no reaction to optokinetic stimulation or a typus inversus, a thing which occurs only if the fixation tonus is very low. In monocular vision with the L.E., on the other hand, the existing jerking nystagmus to L. was considerably enhanced by contour movement to R., while it was abolished by contour movement to L. The pendular nystagmus in early infancy is also indicative of a severe disturbance in the development of the conjugated optomotor reflexes. As regards the reflexes from the L.E., some improvement had taken place in the course of time, at least with respect to the fixation tonus to the right. Here we see once more an asymmetry due to he fact that the *fixation tonus from the better L.E. had developed better than that from the squinting R.E.*

Case 35: J.J., F., 10 yr.

Refraction: bilateral emmetropia. Vis. ac. R.E. 1; L.E. 1; binoc. 1.

No strabismus; no alternating hyperphoria.

Head kept turned to the L.

No strabismus or nystagmus in the family.

Latent nystagmus.

With binocular vision: straight ahead, fine, rather irregular pendular nystagmus with jerks to the L.; to the R., jerking nystagmus to R. with slow frequency and large amplitude; to the L., jerking nystagmus to L. with higher frequency and somewhat smaller amplitude.

In monocular vision with the R.E.: straight ahead, jerks to R.; to the R., jerking nystagmus to R.; to the L., jerks to L.

In monocular vision with the L.E.: straight ahead, fine jerking nystagmus to L.; to the R., eyes practically at rest or very fine jerking nystagmus to R.; to the L., fine jerking nystagmus to L.

In the dark, some wavering of the eyes with a deviation of gaze to the R.

Field of gaze normal.

Good binocular perception and depth perception. Fusion amplitude about 20°.

Optokinetic nystagmus

Binocular with movement to R.: weak jerking nystagmus to L.

„   „   „   „ L.: very weak jerking nystagmus to R.

Monocular R.E. with movement to R.: jerks to L.

„   „ „   „   „   „ L.: weak jerking nystagmus to R.

„   L.E.   „   „   „ R.: „   „   „   „ L.

„   „ „   „   „   „ L.: eyes remained still.

Absence of strabismus, good visual acuity, good stereoscopic vision and normal fusion amplitude: all this points to a normal development of the monocular optomotor reflexes. In contradistinction to this, a very weak fixation tonus (pendular nystagmus) indicates an important disturbance in the development of the conjugated optomotor reflexes. Nevertheless, from each eye a fixation tonus in the direction of the other eye had still developed; this was more marked for the L.E. As a result there was a predominance of the fixation tonus to the R., for which reason the head was turned to the L.

In view of the otherwise good function of the eyes one might be inclined to ask whether the imbalance of the fixation tonus might not perhaps be due in this case to non-optical (e.g. vestibular) reflexes. In our opinion there are strong arguments against this. (1) In view of the latent nystagmus there was certainly a disturbance in the conjugated optomotor reflexes. We do not see how this could be derived from a disturbance in the vestibular reflexes. (2) The terminal-position nystagmus on looking to the R. with the R.E. is difficult to reconcile with a predominance to the R. in the non-optical reflexes. (3) With a predominance of the non-optical gaze tonus to the R. the optokinetic nystagmus with movement of the contours to the R. would have been much more lively. (4) The optokinetic nystagmus in monocular vision with the R.E. also does not fit in with a predominance of a non-optical gaze tonus to the R.

In this case, thus, we come again to the same conclusion as in all the preceding cases of this group: i.e. that the *fixation tonus from one eye, in this case the L.E. was better developed than that from the other eye, as a result of which there was a predominance of the gaze tonus to the R. from the L.E.*

Case 36. J.F.B., M., 47 yr.

Refraction: bilateral myopia: E + 13 D.
Vis. ac. with correction: R.E. $^1/_2$; L.E. $^1/_2$; binoc $^1/_2$.
No strabismus; no alternating hyperphoria.
Myopic changes in the fundus.
Latent nystagmus.

With binocular vision: straight ahead, now and then fine jerks to R.; to the R., jerks to R.; to the L., jerks to L. with rotatory component.

In monocular vision with the R.E.: straight ahead, jerking nystagmus to R.; to the R., jerking nystagmus to R.; to the L., no nystagmus.

In monocular vision with the L.E.: straight ahead, fine jerking nystagmus to L.; to the R., no nystagmus; to the L., jerking nystagmus to L.

Field of gaze normal.

Binocular perception was present but stereoscopic vision was deficient.

Good fusion amplitude.

Optokinetic nystagmus

Binocular with movement to R.: jerking nystagmus to L.
„ „ „ „ L.: „ „ „ R.
Monocular R.E. with movement to R.: no nystagmus.
„ „ „ „ „ „ L.: accentuated jerking nystagmus to R.
„ L.E. „ „ „ R.: „ „ „ „ L.
„ „ „ „ „ „ L.: weak „ „ „ R.

The jerks to R. in binocular vision and the fact that the nystagmus in monocular vision with the R.E. was stronger than that in monocular vision with the L.E. show that there was a predominance of the optical fixation tonus to the L., and also that this predominance must have been due to stimuli from the R.E. This predominance was also manifested in the optokinetic nystagmus with monocular vision. In monocular vision with the R.E. the fixation tonus evoked by the movement of contours to the R. was indeed able to cancel out the existing predominance of the fixation tonus to the L., but not to prevail over it.

In monocular vision with the L.E. the fixation tonus evoked by contour movement to the L. could prevail over the existing predominance of the fixation tonus to the R. It is certain that the difference in fixation tonus for R. and L. turning was greater in monocular vision with the R.E. than in monocular vision wth the L.E. The question is, however, whether this difference was due to a better development of the fixation tonus to the L. from the R.E. or from the L.E. If the function of one eye is very bad it can reasonably be assumed that the optomotor reflexes from this worse eye will have developed less satisfactorily, as in Cases 29, 30, 31, 32, 33, and 34. In Case 35 the pendular nystagmus showed that we had to start from the assumption of a very low fixation tonus for both eyes; this was also manifested by the weak optokinetic nystagmus. In the present case, however, we had a patient with a quite strong reaction to optokinetic stimuli. We are therefore inclined to assume that the predominance of the fixation tonus to the L. was indeed due to the R.E., but that this found expression *not because the fixation tonus from the L.E. had lagged far behind but, on the contrary, because the fixation tonus from the L.E. had developed in a more normal way.*

Case 37. G.D., F., 17 yr.

Refraction: R.E. HAsH $\begin{bmatrix} E - 1\ D. \\ E - 1/2\ D. \end{bmatrix}$ max. 20° nas.;

L.E. H. E — 1 D.

Vis. ac. with correction: R.E.$^2/_5$; L.E. $^2/_5$; binoc. $^3/_5$.
No strabismus. No alternating hyperphoria.
The patient kept her head turned somewhat to the R.

Latent nystagmus only in monocular vision of the R.E.

With binocular vision: straight ahead, eyes at rest; to the R., strong jerking nystagmus to R.; to the L., no jerking nystagmus or a few jerks to L.

In monocular vision with the R.E.: straight ahead, jerking nystagmus to R.; to the R., strong jerking nystagmus to R. with large amplitude; to the L., no nystagmus.

In monocular vision with the L.E.: straight ahead, very weak jerking nystagmus to L.; to the R., no nystagmus; to the L., weak terminal-position nystagmus.

In the dark: eyes at rest most of the time; sometimes jerks to R. and sometimes to L., the former being rather stronger.

Field of gaze normal.

Binocular perception was present, but no depth perception.

Small fusion amplitude and very rapid suppression of retinal images.

Optokinetic nystagmus

Binocular with movement to R.: jerking nystagmus to L.

  „ „ „ „ L.: „   „ „ R.

Monocular R.E. with movement to R.: lively jerking nystagmus to R.

  „ „ „ „ „ „ L.: jerking nystagmus to R., rather less lively.

Monocular L.E. with movement to R.: jerking nystagmus to L.

  „ „ „ „ „ „ L.: irregular movements, sometimes jerks to R., sometimes jerks to L.

The turning of the head to the R., the binocular terminal-position nystagmus to the R. and the optokinetic reaction with monocular vision are all proofs of a predominance of optical fixation tonus to the L. — or perhaps more correctly of a deficiency of optical fixation tonus to the R. The practically unilateral latent nystagmus and also the optokinetic reaction with monocular vision show that this predominance — or this deficiency — was due to the optical fixation tonus supplied by the R.E. The difference in fixation tonus for right and left turning from the R.E. was thus much greater than the difference in fixation tonus for right and left turning from the L.E. As in the previous case, we are here confronted by the question as to whether the fixation reflexes for left turning from the R.E. had developed better or whether the fixation reflexes for both R. and L. turning from the L.E. had developed better. In our opinion the latter is the correct answer, since in binocular vision the fixation reflexes of the L.E. were clearly predominant. In the first place the eyes remained at rest in binocular vision and in the second place there was a practically normal opto-kinetic reaction in both directions with binocular vision. We are therefore of the opinion that in this case also there was *a more balanced development of the optical fixation reflexes from the L.E., so that this eye was the dominant one and the greater asymmetry of the fixation tonus from the R.E. was largely overcome.*

Case 38. v. H., F., 22 yr.

Refraction: bilat. hypermetropia E — 1 D.
Vis. ac. with correction: R.E. 1; L.E. $^2/_5$; binoc. 1.
Strab. conv. oc. sin. No alternating hyperphoria.
With spectacles the position of the eyes was nearly parallel.
Latent nystagmus; stronger with occlusion of the R.E.
Terminal-position nystagmus on looking to R. and to L.
Convergence fairly good.
Binocular perception but insufficient depth perception.

Optokinetic nystagmus

Binocular with movement to R.: no nystagmus.
   „   „   „   „ L.: weak jerking nystagmus to R.
Monocular R.E. with movement to R.: weak   jerking nystagmus to R.
   „   „ „   „   „   „ L.: strong   „   „   „ R.
   „   L.E.   „   „   „ R.: jerking nystagmus to L.
   „   „ „   „   „   „ L.:   „   „   „ L.

With binocular vision there was apparently a predominance of the fixation tonus to L., as demonstrated by the optokinetic nystagmus. Nevertheless, the jerking nystagmus to the L. with monocular vision with the L.E. was stronger than the jerking nystagmus to the R. in monocular vision with the R.E. The resultant of this would rather lead one to expect a predominance of the fixation tonus to the R. with binocular vision. It therefore appears likely that with both eyes open the R.E. predominated to such a degree (master eye) that the stimuli from the L.E. did not get a fair chance (inhibition). The dominance of the R.E. was also accompanied by a better development of the optical fixation reflexes from this eye. This is shown by the effect of optokinetic stimulation with monocular vision, this effect being greater for the good R.E. than for the amblyopic L.E. Here again, thus, the asymmetry was a consequence of the fact that the *fixation tonus from the R.E. had developed better than that from the amblyopic L.E.*

Case 39. W.Z., M., 12 yr.

Refraction: R.E. Hm. E — $1^1/_2$ D.; L.E. Hm. E — 2 D.

Vis. ac. with correction: R.E. $1^1/_2$; L.E. $^3/_5$ (formerly $^6/_{50}$); binoc. $1^1/_2$.

No strabismus. No alternating hyperphoria.

Strabismus in the family.

Latent nystagmus in monocular vision with the L.E. This nystagmus had gradually decreased, together with the amblyopia.

In binocular vision and in monocular vision with the R.E. the eyes remained at rest.

In monocular vision with the L.E.: straight ahead, jerking nystagmus to L.; to the R. no nystagmus; to the L., marked jerking nystagmus to L.

No nystagmus in the dark.

Field of gaze normal.

Binocular perception was present; suppression of the image of the L.E. also occurred frequently.

Fusion amplitude only 8°.

Optokinetic nystagmus

Binocular with movement to R.: jerking nystagmus to L.
    „   „   „   „ L.: eyes at rest.
Monocular R.E. with movement to R.: jerking nystagmus to L.
  „   „ „   „   „   „ L.:   „   „   „ R.
  „   L.E. „   „   „ R.:   „   „   „ L.
  „   „ „   „   „   „ L.: weak jerking nystagmus to L.

The optokinetic nystagmus with binocular vision points to a predominance of the fixation tonus to R. The jerking nystagmus to L. in monocular vision with the L.E. shows that this predominance must originate from the L.E. This case, thus, forms an interesting contrast to the preceding one (Case 38). In the latter case the stimuli from the squinting eye were inhibited in binocular vision. In the present case, with no strabismus, the stimuli from the less good eye also made themselves felt. The fact that the good R.E. remained still in monocular vision and the normal optokinetic nystagmus indicate a practically normal development of the optomotor reflexes from this eye. The history of considerable improvement of the jerking nystagmus in monocular vision with the L.E. suggests that the optomotor reflexes from this eye cannot be so bad either; but still the leftward optical fixation reflexes from the L.E. are insufficient. The asymmetry in the fixation tonus can thus be ascribed to an *insufficient development of the optical fixation reflexes from the worse L.E.,* possibly also related to the amblyopia of this eye.

Case 40. H., M., 29 yr.

Refraction: R.E. Hm. E — 3 D; L.E. Hm, E — 1 D.
Vis. ac. with correction: R.E. $^1/_3$ plus; L.E. 1 with error; binoc. 1.
Slight strab. div. oc. dex.; sometimes alternating.
Strabismus also in the family.
Latent nystagmus; considerably stronger in monocular vision with the R.E.
On looking to the R. and to the L. a few coarse, jerky adjusting movements.
No nystagmus in the dark.
Field of gaze normal.
Alternate vision but no binocular perception. Very poor convergence.

Optokinetic nystagmus

Binocular with movement to R.: weak jerking nystagmus to L.
    „   „   „   „ L.: jerking nystagmus to R.
Monocular R.E. with movement to R.: weak   jerking nystagmus to R.
  „   „ „ „   „   „ L.: strong   „   „   „ R.
  „   L.E. „   „   „ R.: jerking nystagmus to L.
  „   „ „ „   „   „ L.: eyes still; sometimes a few jerks to L.

This patient had no nystagmus in the dark. The optokinetic nystagmus with binocular vision points to a predominance of the optical gaze tonus to L.; this is also very clearly shown by the fact that the nystagmus in monocular vision with the R.E. was much stronger than that with the L.E. The predominance of the gaze tonus was thus of optomotor nature and had its origin in the R.E. The optokinetic nystagmus in monocular vision is also in agreement with the· foregoing. As in Cases 36, 37, 38 and 39 thus, we have here again a *more balanced development of the optical fixation reflexes from the better L.E. and a predominance of optical gaze tonus to L. owing to a more asymmetric development of the optical fixation reflexes from the worse R.E.*

Case 41. H.V.-L., F., 29 yr.

Refraction: R.E. HAsH $\begin{bmatrix} E - \frac{1}{2} D. \\ E - 1\frac{1}{2} D. \end{bmatrix}$ max. $15°$ nas.;

L.E. MAsM. $\begin{bmatrix} E + 2\frac{1}{2} D. \\ E + 2 D. \end{bmatrix}$ max. $50°$ nas.

Vis. ac. with correction: R.E. $\frac{1}{5}$; L.E. $\frac{3}{5}$.

Formerly had strab. conv. oc. dex.; now no strabismus when wearing spectacles.

No alternating hyperphoria.

Congenital cataracta punctata.

Latent nystagmus. No strabismus or nystagmus in the family.

With binocular vision: straight ahead, eyes at rest; to the R. jerking nystagmus to R.; to the L., eyes at rest.

In monocular vision with the R.E.: straight ahead, jerking nystagmus to R.; to the R., strong jerking nystagmus to R.; to the L., jerking nystagmus disappearing.

In monocular vision with the L.E.: straight ahead, jerking nystagmus to L.; to the R., jerking nystagmus disappearing; to the L., jerking nystagmus to L.

In the dark, very weak pendular nystagmus with jerks to R.

Field of gaze normal.

No binocular perception. Insufficient convergence.

Optokinetic nystagmus

Binocular with movement to R.: jerking nystagmus to R.
   ,,   ,,   ::   ,, L.: weak jerking nystagmus to L.
Monocular R.E. with movement to R.: jerking nystagmus to R.
  ,,   ,, ,, ,,   ,,   ,, L.: eyes at rest.
  ,,   L.E.   ,,   ,,   ,, R.: a few jerks to L.
  ,,   ,, ,, ,,   ,,   ,, L.: jerking nystagmus to L.

The analysis and interpretation of the findings in this case was not easy. Some conclusions, however, can be drawn with certainty. In the first place the inverse type of the optokinetic nystagmus, both in binocular and in monocular vision, shows that the fixation tonus must have been very low. In the second place there was, in binocular vision, a predominance of the fixation tonus to the L. This is shown by the strong terminal-position nystagmus on looking to the R., and further also by the optokinetic nystagmus with binocular vision, since the inverse type is as a rule stronger the weaker is the fixation tonus in the direction of the moving contours. In the third place we can deduce from the terminal-position nystagmus in monocular vision with the R.E. and with the L.E. that the difference in fixation tonus for right and left turning was greater in monocular vision with the R.E. But we cannot assume that the optical fixation reflexes had developed better from the worse R.E. than from the better L.E. We thus come to the conclusion that the difference in fixation tonus for right and left turning from the R.E. was indeed greater, but the fixation tonus from the better L.E. was at a somewhat higher level for both directions. In agreement with this is the fact that with monocular vision a nasalward movement of the contours gave a stronger inverse type with the R.E. than with the L.E. By inverse type we understand in this case also the weakening instead of strengthening of the existing jerking nystagmus. Thus we come to a satisfactory explanation, viz: that here again the asymmetry is to be ascribed to a *somewhat better development of the optical fixation reflexes from the eye with the better visual acuity.* The inverse type excludes an explanation based on a predominance due to non-optical reflexes.

Summary of Group IV.

This group comprised 13 patients with latent nystagmus and an asymmetric optical gaze tonus. This asymmetry was often manifested by fine jerks in a given direction with binocular vision, by a difference in terminal-position nystagmus in the 2 directions, by a difference in jerking nystagmus in monocular vision with the R.E. and with the L.E. and by the optokinetic nystagmus in both binocular and monocular vision. There was a striking correspondence among all these phenomena. In all our patients of this group the latent nystagmus was due to predominance of the fixation tonus with monocular vision in the direction of the covered eye.

In Case 34 there had been a pendular nystagmus in childhood; this had become transformed into a latent nystagmus, i.e. a fixation tonus had developed which was oppositely directed for each of the eyes.

The difference in fixation tonus for right and left turning that is thus established is not necessarily the same for both eyes. In Case 34 this was also not the case. The difference in fixation tonus was stronger in the better L.E. than in the worse R.E. An unequal development of this kind, with a predominance of the nasalward fixation tonus in monocular vision with the better eye was also seen, in varying degrees, in Cases 29, 30, 31, 32 and 33.

In Cases 35 and 36 it was not possible to speak of a better or worse eye, and yet they showed a definite asymmetry. While the strength of the jerking nystagmus in monocular vision is dependent on the *difference* between the fixation tonus for right turning and for left turning, it does not depend on the *level* of the fixation tonus. Thus we come to the conclusion that in Cases 35 and 36 the difference of the fixation tonus was indeed greater for one of the eyes, while the fixation tonus for the other eye had developed more equally in both directions; in Case 35 at a lower level and in Case 36 at a higher level. In Cases 37, 38, 39, 40 and 41 the difference in fixation tonus was greater in monocular vision with the worse eye and the fixation tonus for right and left turning had developed more equally in the better eye: in Case 39 at a very high level and in Case 41 at a very low level. Case 38 had in binocular vision an inhibition of the optomotor reflexes from the worse eye with the greater difference in fixation tonus, as a result of which the fixation tonus from the better eye predominated in binocular vision.

Group V. Patients with latent nystagmus and asymmetric non-optical and optical gaze tonus.

In our investigation of cases of latent nystagmus we found 6 patients who showed jerking nystagmus in the dark whereas there was little or no nystagmus in the light. This pointed to a unilaterally directed predominance of the non-optical gaze tonus. Oddly enough, all these 6 patients had a definitely weaker nystagmus in monocular vision with the better or preferred eye than with the worse eye. This suggests that the *non-optical* predominance of gaze tonus had developed gradually as a greater or less degree of compensation of a predominance of *optical* gaze tonus in the opposite direction. This led us to investigate, in another patient who also showed a stronger jerking nystagmus in monocular vision with the worse eye, whether all the phenomena of her case could also be explained by the assumption that an asymmetric non-optical gaze tonus was present. Thus Group V comprises 7 cases of latent nystagmus in which we believe that an asymmetric non-optical gaze tonus was present as well as an asymmetric optical gaze tonus.

Case 42. L.T., M., 22 yr.

Refraction: R.E. HAsH. $\left[\begin{array}{l} E - 2\,D. \\ E - 2^3/_4\,D. \end{array}\right.$ max. 40° nas.;

L.E. HAsH. $\left[\begin{array}{l} E - {}^1/_2\,D. \\ E - 2\,D. \end{array}\right.$ max. vertic.

Vis. ac. with correction: R.E. 1; L.E. $<$ $^1/_{60}$; binoc. 1.

Strab. conv. oc. sin.; slight strab. sursoadductorius suggestive of alternating hyperphoria.

Strabismus also in the family.

Latent nystagmus; formerly also rotatory nystagmus of the R.E.

With binocular vision sometimes slight jerking nystagmus to L.

In monocular vision with the R.E. weak jerking nystagmus to R.

In monocular vision with the L.E. a very strong jerking nystagmus to L.

In the dark a definite jerking nystagmus to L.

Slight limitation of adduction of the L.E. (following a strabismus operation); field of gaze further normal.

Optokinetic nystagmus

Binocular with movement to R.: jerking nystagmus to L.

„   „   „   „ L.:   „   „   „ R.

Monocular R.E. with movement to R.:   weak, irregular jerking nystagmus
                                                               to R.

„   „ „   „   „   „ L.: jerking nystagmus to R.

„   L.E.   „   „   „ R.: strong jerking nystagmus to L.

„   „ „   „   „   „ L.: jerking nystagmus to L.

The strong jerking nystagmus to L. in the dark can only be explained by a predominance of the non-optical gaze tonus to the R. This predominance of the non-optical gaze tonus to the R. practically balanced, with both eyes open, an excess of optical gaze tonus to the L. this latter being the resultant of the optomotor reflexes from both eyes. In monocular vision with the

R.E. the excess of non-optical gaze tonus to the R. was not in itself capable of overcoming the excess of optical gaze tonus to the L. from the R.E. In order for this to be overcome the aid of the excess of optical fixation tonus to the R. from the L.E. was necessary as well; from time to time this even gave rise to a slight jerking nystagmus to the L. The optokinetic nystagmus, both with binocular and wit monocular vision, was also in agreement with this idea. In this case, thus, *the excess of non-optical gaze tonus had adapted itself to the resultant of the optical fixation tonus from both eyes.*

Case 43. W. v. d. B., F., 10 yr.

Refraction: bilat. HAsH. $\left[\begin{array}{l} E - 5 \text{ D.} \\ E - 6 \text{ D.} \end{array} \right.$

Vis. ac. with correction: bilat. 1; binoc. 1.
Strab. conv. alt. from the first year of life.
Left eye preferred.
Alternating hyperphoria.
Latent nystagmus; strong jerking nystagmus to R. when the R.E. fixated; weak jerking nystagmus to L. when the L.E. fixated.
In the dark a weak jerking nystagmus to R., sometimes with endorotation of the L.E.; occasionally fine jerks to the L. and then with endorotation of the R.E.; the R.E. then often stopped moving.
Field of gaze normal.
No binocular perception.

Optokinetic nystagmus

Binocular with movement to R.: slight jerking nystagmus to L.
   „   „   „   „ L.:   „   „   „   „ R.
Monocular R.E. with movement to R.: weak   jerking nystagmus to R.
   „   „ „   „   „   „ L.: strong   „   „   „ R.
   „   L.E.   „   „   „ R.: jerking nystagmus to L.
   „   „ „   „   „   „ L.: rather irregular pendular nystagm.

The difference in gaze tonus for right and left turning in monocular vision with the L.E. was undoubtedly smaller than that in monocular vision with the R.E. The nystagmus in the dark indicates a predominance of the non-optical gaze tonus to the L. This case greatly resembled the previous one (Case 42); as far as the nystagmus was concerned these 2 cases were each other's perfect mirror images. Here again there was a *slight predominance of non-optical gaze tonus to the L. as a partial compensation of a predominance of optical fixating tonus to the R. from the L. (preferred) eye. The non-optical gaze tonus had counterbalanced the resultant of the optical gaze tonus from both eyes.*

Case 44. F.J.A.T.P., M., 23 yr.

Refraction: R.E. M. E + 1 D.; L.E. M. E + 3 D.
Vis. ac. with correction: R.E. $^1/_{10}$; L.E. $^1/_4$; binoc. $^1/_4$.
No strabismus. No alternating hyperphoria.
Albinotic fundi.
Pendular nystagmus and latent nystagmus of the R.E.
Nystagmus stated to have been present from birth but to have improved greatly.
With binocular vision: straight ahead, pendular nystagmus alternating with jerks to R.; to the R., jerking nystagmus to R.; to the L., jerking nystagmus to L.
In monocular vision with the R.E.: straight ahead, marked jerking nystagmus to R.; to the R., increased jerking nystagmus; to the L., decreased jerking nystagmus.
In monocular vision with the L.E.: straight ahead, pendular nystagmus with now and then jerks to R. and sometimes also to L.; to the R., jerking nystagmus to R.; to the L., jerking nystagmus to L.
In the dark, definite jerking nystagmus to R.
Field of gaze normal.
Binocular perception and also depth perception were present, but failure occurred with difficult stereoscopic plates. Fusion amplitude about 12°.
Normal cortical binocular junction as tested with after-images.

Optokinetic nystagmus

Binocular with movement to R.: weak   jerking nystagmus to L.
   ,,   ,,   ,,   ,, L.: strong   ,,   ,,   ,, R.
Monocular R.E. with movement to R.: weak   jerking nystagmus to R.
   ,,   ,, ,,   ,,   ,,   ,, L.: strong   ,,   ,,   ,, R.
   ,,   L.E.   ,,   ,,   ,, R.: weak   ,,   ,,   ,, L.
   ,,   ,, ,,   ,,   ,,   ,, L.:   ,,   ,,   ,,   ,, R.

There was an unmistakable predominance of the non-optical gaze tonus to L. in this case: the jerking nystagmus to R. in the dark was much stronger than in daylight. In daylight this was apparently counterbalanced by the optical reflexes from the better L.E. The non-optical excess of the gaze tonus showed up but little in the terminal-position nystagmus in this case. This excess was, however, recognizable in the optokinetic nystagmus, both in binocular and in monocular vision. Thus we find that with monocular vision with the R.E. and movement of the contours to the L. a much stronger jerking nystagmus was produced than with movement of the contours to the R., while in monocular vision with the L.E. a movement of the contours to the L. was able to change the existing pendular nystagmus into a jerking nystagmus to the R. and a contour movement to the R. evoked only a weak jerking nystagmus to the L. We therefore conclude that *an excess of non-optical gaze tonus to L. had developed as a compensation of the excess of optical fixation tonus to the R. from the dominant L.E.* In this case, thus, there was not a compensation of the resultant of the fixation tonus from both eyes but only a compensation of the fixation tonus from the master eye. This is also shown by the jerks to the R. in binocular vision.

Case 45. H. d. R., F., 23 yr.

Refraction: R.E. ?; L.E. HAsH. $\left[\begin{array}{l} \text{E} - \frac{1}{2}\,\text{D.} \\ \text{E} - 1\,\text{D.} \end{array} \right.$ max. vertic.

Vis. ac. with correction: R.E. $^5/_{60}$; L.E. 1; binoc. 1.
Strab. div. oc. dex. from birth.
Strabismus also in the family.
Alternating hyperphoria.
Rotatory pendular nystagmus and latent nystagmus of the R.E.
With binocular vision to the R. and to the L. the rotatory pendular nystagmus remained.
In monocular vision with the R.E.: straight ahead, rotatory pendular nystagmus with jerking nystagmus to R.; to the R., rotatory jerking nystagmus to R.; to the L., horizontal jerking nystagmus to R.
In monocular vision with the L.E.: rotatory pendular nystagmus in all directions, sometimes a few jerks to L.
In the dark, jerking nystagmus to R.
Field of gaze normal.
No binocular perception.

Optokinetic nystagmus

Binocular with movement to R.: weak jerking nystagmus to L.

„   „   „   „ L.: jerking nystagmus to R.

Monocular R.E. with movement to R.: weak   jerking nystagmus to R.

„   „ „   „   „   „ L.: marked   „   „   „ R.

„   L.E.   „   „   „ R.: eyes at rest.

„   „ „   „   „   „ L.: lively jerking nystagmus to R.

In this case, thus, there was a jerking nystagmus to R. in the dark that was absent in daylight. There must therefore have been a predominance of the non-optical gaze tonus to the L. As no jerking nystagmus occurred in daylight, there must have been a counterbalancing excess of the optical gaze tonus to the R. This originated, as shown by the findings with monocular vision, from the L.E. The optical gaze tonus from the R.E. and the L.E. thus corresponded perfectly to what would be expected from a latent nystagmus, but it was complicated by an excess of the non-optical gaze tonus to the L. It goes without saying that under these conditions a contour movement to the L. will, with binocular vision, give a stronger optokinetic nystagmus than will a contour movement to the R. In monocular vision with the worse R.E. the excess of gaze tonus to the L. was supplied by the optical tonus from this eye plus the excess of the non optical gaze tonus, so that there is nothing surprising in the fact that the jerking nystagmus to the R. could not be overcome by optokinetic stimulation. In monocular vision with the better L.E., a contour movement in the direction of the non-optical excess had more effect than a contour movement in the direction of the optical excess. This is the general rule, and more especially so where the optical fixation tonus in general is rather low (pendular nystagmus). We therefore consider that there are good grounds for the assumption that in this case, as in the previous one (Case 44), there was a *predominance of the non-optical gaze tonus in one direction as a compensation for an excess of optical gaze tonus in the opposite direction, originating from the better eye.*

Case 46. J.N., M., 17 yr.

Refraction: R.E. AsH $\begin{bmatrix} E \\ E - 2^1/_2\ D. \end{bmatrix}$ max. $30°$ nas.;

L.E. AsH. $\begin{bmatrix} E \\ E - 1/_2\ D. \end{bmatrix}$ max. $10°$ temp.

Vis. ac. with correction: R.E. $^1/_5$; L.E. $^6/_5$; binoc. $^6/_5$.

Strab. conv. oc. dex. Alternating hyperphoria.

The R.E. showed residual effects of an iris prolapse due to an injury at the age of 5 yr.

Latent nystagmus of the R.E.

Terminal position nystagmus on looking to the R. and to the L.; somewhat less strong on looking to the L.

In the dark, jerking nystagmus to the R.

Sometimes simultaneous perception with both eyes but no binocular fusion and no depth perception. The retinal image of the R.E. was very readily suppressed, especially the central part.

Optokinetic nystagmus

Binocular with movement to R.: jerking nystagmus to L.

„ „ „ „ L.: „ „ „ R.

Monocular R.E. with movement to R.: weak, rather irregular jerking nystagmus to R.

„ „ „ „ „ „ L.: strong jerking nystagmus to R.

„ L.E. „ „ „ R.: jerking nystagmus to L.

„ „ „ „ „ „ L.: „ „ „ R.

This case was a replica of the 2 preceding ones (Cases 44 and 45), except that perhaps the predominance of the non-optical gaze tonus to the L. was here rather less pronounced, in consequence of which it did not show up so well in the optokinetic nystagmus with binocular vision. The optokinetic nystagmus with monocular vision was practically identical with that in Case 44. Therefore we only wish to repeat that here again we have *a predominance of non-optical gaze tonus to the L. as a compensation for the predominance of optical fixation tonus to the R. supplied by the better L.E.*

Case 47. H.J.D., M., 27 yr.

Refraction: R.E. MAsM. $\begin{bmatrix} E + 6\,D. \\ E + 5\,D. \end{bmatrix}$ max. 20° temp.;

L.E. MAsM. $\begin{bmatrix} E + 6\,D. \\ E + 5\,D. \end{bmatrix}$ max. 10° temp.

Vis. ac. with correction: R.E. $^2/_5$; L.E. $^1/_5$ plus.

Strab. div. oc. sin. No alternating hyperphoria.

Latent nystagmus of the L.E. with rotatory component.

With binocular vision: straight ahead, very slight jerking

nystagmus to L.; to the R., eyes at rest; to the L., jerking nystagmus to L.

In monocular vision with the R.E.: straight ahead, eyes at rest; to the R., very slight jerking nystagmus to R.; to the L., eyes at rest.

In monocular vision with the L.E.: straight ahead, jerking nystagmus to L.; to the R., eyes at rest; to the L., strong jerking nystagmus to L.

In the dark, jerking nystagmus to L.

Field of gaze normal.

No binocular perception.

Optokinetic nystagmus

Binocular with movement to R.: jerking nystagmus to L.

    „   „   „   „ L.:   „   „   „ L.

Monocular R.E. with movement to R.: eyes at rest.

    „   „ „   „   „   „ L.: jerks to R.

    „   L.E.   „   „   „ R.: marked jerking nystagmus to L.

    „   „ „   „   „   „ L.: jerking nystagmus to L.

The jerking nystagmus to L. observed in the dark, only a trace of which remained in daylight, points to a predominance of the non-optical gaze tonus to the R. In monocular vision with the R.E. this was compensated by an excess of optical gaze tonus to the L. Here we have once more the usual relationships of the optical fixation tonus in latent nystagmus, complicated by an excess of non-optical gaze tonus in a given direction. The nystagmus on looking to the R. and to the L. requires no further explanation; the same applies to the optokinetic nystagmus with binocular vision and contour movement to the R. A rather different matter is the jerking nystagmus to L. (inverse type) with movement of the contours to the L. Apparently the non-optical gaze tonus here retained the upper hand, while the optical gaze tonus became exhausted. With monocular vision with the R.E. and contour movement to the R. the optokinetic stimulation had practically no effect ('optische Drehstarre'). With monocular vision with the R.E. and contour movement to the L. the optokinetic stimulation prevailed despite the non-optical predominance to the R. In monocular vision with the L.E. the jerking nystagmus to the L. resulted from the excess of optical gaze tonus to the R. plus the non-optical gaze tonus to the R.; it is understandable that this jerking nystagmus to the L. could not be overcome by optokinetic stimuli. In this case also we are of the opinion that there was an *excess of non-optical gaze tonus in one direction as a compensation for an excess of optical gaze tonus in the opposite direction, originating from the better eye.*

Case 48. W.J., F., 27 yr.

Refraction: R.E. MAsM $\begin{bmatrix} E + 5^1/_2 \text{ D.} \\ E + 3.5 \text{ D.} \end{bmatrix}$ max. 30° temp.;

L.E. MAsM $\begin{bmatrix} E + 2^1/_2 \text{ D.} \\ E \end{bmatrix}$ max. 40° nas.

Vis. ac. with correction: R.E. $^3/_5$; L.E. $^1/_{10}$ plus; binoc. $^3/_5$. Strab. div. et sursoabductorius oc. sin.

R.E. coloboma iridis; L.E. coloboma iridis et chorioideae with central scotoma.

Latent nystagmus of the L.E.

On looking to the R., eyes at rest; on looking to L., slight jerking nystagmus to L.

Optokinetic nystagmus

Binocular with movement to R.: jerking nystagmus to L.
     „   „   „   „ L.:   „   „   „ L.
Monocular R.E. with movement to R.: slight jerking nystagmus to L.
     „   „ „   „   „   „ L.:   „   „   „   „ L.
     „   L.E.   „   „   „ R.: strong   „   „   „ L.
     „   „ „   „   „   „ L.:   „   „   „   „ L.

This patient was not examined for nystagmus in the dark. We are therefore confronted with the question of whether the fact that the eyes remained at rest in monocular vision with the better R.E. was a consequence of more equal development of the optical reflexes from this eye or whether the eyes were kept still by opposed predominance in the non-optical and the optical reflexes. The latter was almost certainly the case. With an equal development of the optical fixation reflexes the optokinetic stimulation of the R.E. would not have evoked a normal optokinetic nystagmus in one direction and an inverse type in the other direction. Under conditions of asymmetric gaze tonus, however, this is not infrequent, as the non-optical gaze tonus persists and the optical gaze tonus may suffer exhaustion or inhibition. We are therefore led to assume that there was a predominance of non-optical gaze tonus to the R. This also accounts for the strong jerking nystagmus to L. in monocular vision with the L.E., as the predominance was then caused both by the non-optical gaze tonus and by the optical gaze tonus from the L.E.

Thus, even without the confirmation of nystagmus in the dark[1]) we feel justified in assuming here also that *an excess of non-optical gaze tonus to the R. had developed to compensate an excess of optical fixation tonus to the L. from the better R.E.*

Summary of Group V.

In the 7 patients of this group the jerking nystagmus in monocular vision with the one eye was very marked, while in monocular vision with the other eye it was very weak or entirely absent. In 6 cases the jerking nystagmus was stronger when the worse eye was used. In Case 43 the visual acuities of both eyes were the same, but here also the jerking nystagmus was weaker when the preferred eye was used. In 6 cases there was a marked jerking nystagmus in the dark. In Case 48 this point was not investigated. Nystagmus in the dark indicates a predominance of the non-optical gaze tonus in a given direction. This predominance is regarded as a more or less complete compensation of an excess of optical fixation tonus in the opposite direction, originating from the better or preferred eye. In 2 patients (Cases 42 and 43) the excess of non-optical gaze tonus balanced the resultant of the gaze tonus from the 2 eyes. The excess of non-optical gaze tonus is always manifested in the optokinetic nystagmus, but it is not always manifested in the terminal position nystagmus, as demonstrated by Cases 44 and 45. In Case 44 we saw a jerky terminal-position nystagmus on looking both to the R. and to the L.; in Case 45 the rotatory pendular nystagmus persisted unchanged on looking in either direction.

Group VI. One-eyed patients with latent nystagmus.

This group comprised 7 one-eyed persons who showed when their remaining eye was covered — and thus also in the dark — a jerking nystagmus with the fast phase in the direction of the blind eye. Actually these 7 patients represent limiting cases of the preceding group (Gr. V).

If, as in latent nystagmus, there is with monocular vision a predominance of fixation tonus in the nasalward direction and

---

[1]) This patient recently visited our consulting-room again and it was then found that she did indeed have a jerking nystagmus to L. in the dark.

this is compensated by a non-optical gaze tonus in the opposite direction, then in one-eyed subjects the jerking nystagmus would be expected to disappear gradually in daylight and to appear in the dark, but then with the fast phase in the direction of the blind eye. This deduction was fully confirmed by the phenomena observed in the following cases.

Case 49: H.R., M., 17 yr.

Refraction: R.E. M. E + ¹/₂ D. Vis. ac. with correction: R.E. 2.

On the left the patient wore an artificial eye; the L.E. had been enucleated at the age of 6 months on account of pseudoglioma.

When covered the R.E. frequently turned somewhat upwards.

On looking straight ahead no movement or sometimes a few irregular, fine movements.

On looking to the R. jerking nystagmus to R.; on looking to the L. pendular nystagmus with a few jerks to L.

When the R.E. was covered, jerking nystagmus to L. if the gaze was now turned to the R., jerking nystagmus to R. occurred; with gaze to the L. jerking nystagmus to L.

In the dark also jerking nystagmus to L., with concomitant movement of the artificial eye.

Optokinetic nystagmus:

Monocular R.E. with movement to R.: eyes first still for a moment, then coarse jerks to L.

Monocular R.E. with movement to L.: eyes still at first; then small jerks to L.

The jerking nystagmus to L in the dark is evidence of a predominance of non-optical gaze tonus to R. The disappearance of the nystagmus in daylight is evidence of an opposed predominance of optical fixation tonus to the L. The terminal-position nystagmus on looking to the R. and the pendular nystagmus on looking to the L. with the eye open show that the optical fixation tonus to the R. was the more deficient. When the excess of optical gaze tonus to the L. was eliminated by occlusion of the eye, direction of the gaze to the L. also gave a jerking nystagmus to L. The concomitant movement of the prosthesis shows that we are dealing here with a reflex to conjugated movement. The optokinetic nystagmus was in agreement with a predominance of non optical gaze tonus to the R. With contour movement to the R. the excess of optical fixation tonus to the L. was inhibited; with contour movement to the L. the excess of non-optical gaze tonus to the R. could not be inhibited. A fact worthy of note is that with contour movement to the L. the optical fixation tonus to the L. seemed to become gradually exhausted. The phenomena observed in this case do not tell us whether the excess of non-optical or of optical gaze tonus was primary, but on the grounds of observations in other cases we feel justified in concluding that an *excess of non-optical gaze tonus to the R. had compensated for the deficiency of optical fixation tonus to the R.*

Case 50: E.H., F., 31 yr.

Refraction: R.E. AsH. $\begin{bmatrix} E \\ E - \frac{1}{2} D. \end{bmatrix}$

Vis. ac. with correction: R.E. 1.

On the left the patient wore an artificial eye. She gave a history of a microphthalmus on the L., which had been removed at the age of 3 yr.

No strabismus or nystagmus in the family.

On looking straight ahead; eyes sometimes at rest; sometimes rotatory pendular nystagmus; now and than rotatory jerks to R.; on looking to the R., rotatory jerking nystagmus to R.; on looking to the L., diagonally-directed pendular nystagmus, sometimes rotatory jerks to R.

On covering of the R.E. and in the dark, jerking nystagmus to L., with concomitant movement of the artificial eye.

Field of gaze of the R.E. normal.

Optokinetic nystagmus:

Monocular R.E. with movement to R.: jerking nystagmus to L.

    „   „ „   „   „   „ L.: pendular nystagmus; sometimes also jerking nystagmus to L. upwards.

This case was in all respects a perfect replica of the previous one (Case 49), so that further discussion is superfluous and we may conclude here also that an *excess of non-optical gaze tonus to R. had compensated for a deficiency of optical fixation tonus to the R.*

Case 51: W.A., F., 8 yr.

Refraction: L.E. H. E — 3 D. Vis. ac. with correction: L.E. 1. The R.E. was a microphthalmus with congenital cataract and

posterior synechiae; its vision was limited to light perception
($^1/\infty$).

Strab. conv. oc. dex. first noticed at the age of 4 months.

At the time of examination, on looking straight ahead a highly
variable adduction of the R.E.; sometimes the eyes took up a
parallel position.

Upon looking straight ahead and to the R. the L.E. was at
rest; on looking to L. there was jerking nystagmus to L.

With the L.E. covered: jerking nystagmus to R. in all directions of gaze.

In the dark, jerking nystagmus to R.

Field of gaze of the L.E. normal.

Optokinetic nystagmus:

Monocular L.E. with movement to R.: weak   jerking nystagmus to L.
  „   „ „   „   „   „ L.: strong   „   „   „ R.

It goes practically without saying that with the minimal visual
acuity of the R.E. it made no difference whether the L.E. only
was covered or the patient placed in the dark. The fact that a
contour movement to the R. gave still a weak jerking nystagmus
to L. shows that the predominance of the non-optical gaze tonus
to the L. did not exert so great an influence as the predominance
to R. in the 2 previous cases. Apart from this, however, this child
presented us with a mirror image of the 2 previous cases (Cases
49 and 50), so that we may conclude that *an excess of non-
optical gaze tonus to the L. had compensated for the deficiency
of optical fixation tonus to the L.*

Case 52: M.A., F., 17 yr.

Refraction: R.E. H. E — $2^1/2$ D. Vis. ac. with correction
R.E. $^3/_5$.

The L.E. was a microphthalmus with congenital central cataract and anterior synechiae of the stroma iridis. Epicanthi.

Vision of the L.E. limited to light perception ($^1/\infty$).

Strab. div. oc. sin.

In daylight and on looking straight ahead, fine tremor of the
L.E.

With the R.E. covered, strong jerking nystagmus to L.

The same with both eyes covered.

Jerking nystagmus to L. also occurred when a spherical lens of
+ 20 was placed before the R.E.

Field of gaze normal.

Monocular R.E. with movement to R.: very strong jerking nystagmus to L.

„ „ „ „ „ „ L.: weak jerking nystagmus to R.

In this case it appeared that a lens of sph. $+$ 20 was capable of evoking the jerking nystagmus. This shows that the fixation tonus was then eliminated; this fixation tonus was thus only formed when contours were represented on the retina, so that it must not be confused with the light tonus. Apart from this the phenomena in this case were so exactly similar to those of the previous cases that we may conclude that an *excess of non-optical gaze tonus to the R. had compensated for the deficiency of optical fixation tonus to the R.*

Case 53: J.O., M., 36 yr.

Refraction: L.E. emmetropic. Vis. ac.: R.E. $^1/\infty$; L.E. 1.

The patient had undergone an operation for congenital cataract at the age of 30 yr. and it was then found that this eye had an old detachment of the retina. The cataract had been first observed at the age of 2 yr. There was no strabismus horizontalis. The R.E. turned somewhat upward in abduction.

When the L.E. was covered the R.E. turned somewhat downwards; as a rule the R.E. was somewhat higher. There was thus a suggestion of alternating hyperphoria.

On looking straight ahead, slight jerking nystagmus to L. — in the R.E. with a downward vertical component; on looking to the L. the jerking nystagmus to L. became stronger.

When the L.E. was covered, lively jerking nystagmus to R.; on looking to the R. with te L.E. covered the jerking nystagmus to the R. increased; on looking to the L. it decreased.

In the dark also a lively jerking nystagmus to R.

Optokinetic nystagmus:

Monocular L.E. with movement to R.: first normal jerking nystagmus to L; gradual cessation of movement and finally a few jerks to R.

„ „ „ „ „ „ L.: jerking nystagmus to R.; gradually increasing.

The characteristics of alternating hyperphoria and the slight jerking nystagmus to L. in daylight make it highly probable that the excess of optical fixation tonus to the R. was primary.

A remarkable feature was also the course of the optokinetic nystagmus, especially with movement of the contours to the R. where the gradual exhaustion of the optical fixation tonus was so clearly demonstrated. This case is an excellent example of an *excess of non-optical gaze tonus to the L. which had almost completely compensated a deficiency of optical fixation tonus to the L.*

Case 54. O.O., M., 32 yr.

Refraction: L.E. HAsH. $\begin{bmatrix} E - 3\,D. \\ E - 4\,D. \end{bmatrix}$ max. vertic.

Vis. ac. with correction: L.E. $^3/_4$ plus.

R.E.: posterior cortical cataract; detachment of the retina; pigment streaks in the retina. The R.E. was totally blind. The L.E. was dichromat.

Strab. div. oc. dex. No alternating hyperphoria.

The head was held somewhat inclined towards the left shoulder.

Nystagmus had been noticed at a very early age.

No strabismus or nystagmus in the family.

In daylight and looking straight ahead there was no nystagmus as a rule; sometimes slight movement of the eyes with now and then a few jerks to the R. or L.

On looking to the R. or to the L., marked terminal position nystagmus with high frequency and relatively small amplitude.

On looking upward or downward, practically the same as on looking straight ahead.

With the L.E. covered, jerking nystagmus to R. in all directions of gaze but least on looking to the L.; the same in the dark.

Field of gaze of the L.E. normal.

Optokinetic nystagmus:

Monocular L.E. with movement to R.: first pendular nystagmus; then occasional jerks to R.

„ „ „ „ „ „ L.: normal jerking nystagmus to R. with small amplitude.

The jerking nystagmus to R. when the L.E. was covered points to an asymmetry of the non-optical tonic innervation with a predominance to the L. With the eye open and looking straight ahead this non-optical excess to the L. was apparently compensated by an optical excess to the R. There was thus also

an asymmetry of the optical tonic innervation. The optical fixation tonus, however, was insufficiently developed both to the L.
and to the R., as may be concluded from the jerky terminal-
position nystagmus in both directions. This also makes it highly
probable that the *asymmetry in the development of the optical
fixation reflexes was primary and the asymmetry of the non-
optical gaze tonus secondary.* The mediocre visual acuity suggests that the monocular optomotor reflexes had also been somewhat backward in development.

Case 55: H.D.W., M., 20 yr.

Refraction: L.E. HAsH. $\begin{bmatrix} E - 2\,D. \\ E - 5^1/_2\,D. \end{bmatrix}$ max. $56°$ temp.

Vis. ac. with correction: L.E. 1 plus.

The R.E. showed maculae corneae and anterior synechiae; this
had been noticed 4 days after birth. Its vision was limited to
the perception of hand movements at a distance of $^1/_2$ metre
($^{0.5}/_{300}$).

Slight strab. convergens, practically from birth.

No strabismus in the family.

On looking straight ahead the eyes remained still most of
the time; sometimes slight jerking nystagmus to L.; on looking
to the R. the eyes were quite motionless; on looking to the L.
there was an appreciable increase of the jerking nystagmus to L.

With the L.E. covered, jerking nystagmus to R. in all directions of gaze; when an attempt was now made to direct the
gaze to the R., a very strong jerking nystagmus to R. appeared.
When the L.E. was covered the R.E. went into adduction.

Covering of the R.E. had no effect, except that the R.E. turned slightly upward with exorotation and when uncovered turned
somewhat downwards again with endorotation. This was reminiscent of the phenomena of alternating hyperphoria.

In the dark, jerking nystagmus to R.

Field of gaze normal.

Optokinetic nystagmus:

Monocular L.E. with movement to R.: very weak jerking nystagmus to L.
     „   „ „   „   „   „ L.: marked jerking nystagmus to R.

Covering and uncovering of the R.E. made no difference to
the optokinetic nystagmus.

The suggestion of an alternating hyperphoria was seen also

in Case 53. This strongly suggests that the predominance of the optical fixation tonus to the R. must have been primary. The jerking nystagmus to L. on looking straight ahead, although only weak, was also in favour of this view.

As this case corresponded in all respects to the preceding cases we assume that here also an *excess of non-optical gaze tonus to the L. had compensated for the deficiency of optical fixation tonus to the L.*

Summary of Group VI.

This group comprised 7 patients all of whom had only one seeing eye. Occlusion of this eye gave a conjugated jerking nystagmus in the contralateral direction. This signified an excess of non-optical gaze tonus in the homolateral direction. In daylight, therefore, there must have been an excess of optical fixation tonus in the contralateral or a deficiency in the homolateral direction to balance the non-optical gaze tonus. The asymmetry in the optical fixation tonus corresponded to that in latent nystagmus. Since there was further in 2 patients (Cases 53 and 55) a definite suggestion of alternating hyperphoria, while Case 54 certainly had a deficient development of the optical fixation reflexes, we are of the opinion that the deficiency of optical fixation tonus in the temporalward direction was primary.

As we have given a summary at the end of each group, it is unnecessary to repeat everything here for the whole series. We shall therefore content ourselves with the following summing-up:

The first group comprised 8 patients with pendular nystagmus which could be ascribed to a deficiency of the optical gaze tonus.

The 2nd group comprised 5 patients with pendular nystagmus and an asymmetric optical gaze tonus. It appeared that both a disturbance in the development of the conjugated optomotor reflexes and a disturbance in the monocular optomotor reflexes can cause such asymmetry.

The 3rd group consisted of 15 patents with latent nystagmus and a symmetrical gaze tonus with both eyes open. Latent nystagmus is due to a disturbance in the development of the conjugated optomotor reflexes. The degree of such disturbance could be judged from the greater or less intensity of the terminal-position nystagmus and the greater or less degree of abnormality of the optokinetic nystagmus. In many of these cases a disturbance in the development of the monocular optomotor reflexes was also encountered, but this was not proportional to the severity of the latent nystagmus.

The 4th group comprised 13 patients with latent nystagmus and an asymmetric optical gaze tonus. In 7 cases this asymmetry was due to a better development of the nasally directed fixation tonus from the better eye, so that this better eye showed the stronger jerking nystagmus in monocular vision. In 6 cases the fixation tonus from the better eye appeared to have developed more equally, so that the stronger jerking nystagmus was seen in monocular vision with the worse eye.

The 5th group consisted of 7 patients with latent nystagmus and an asymmetric gaze tonus based on non-optical reflexes, which led in all cases to jerking nystagmus in the dark. It appeared probable that this asymmetry had developed secondarily to compensate an asymmetry in the optical fixation tonus, either from both eyes together or from the more used and better eye.

The 6th group comprised 7 patients with one blind or practically blind eye. In the dark or when the good eye was covered these patients showed a jerking nystagmus in the direction of the blind eye. Here also there was an asymmetry of the non-optical gaze tonus as a compensation for the primary asymmetry in the optical gaze tonus.

DISCUSSION

In the preceding chapter we have attempted, for each case separately, to draw a conclusion from certain phenomena as to the direct cause of the nystagmus. In this chapter we shall first study the manner in which the anomalies and manifestations observed in the individual cases are distributed over the series as a whole, in order to reach a better understanding of the relationship between these phenomena and the nystagmus. Making use of the insight thus achieved we shall then endeavour to ascertain whether we can place our theoretical views upon a sounder basis and whether we can find promising lines of approach to problems in which our present knowledge is insufficient.

For the sake of clarity we have divided our subject-matter into the following sections:

    I. Refraction and visual acuity.
   II. Strabismus and alternating hyperphoria.
  III. Nystagmus on looking sideways.
   IV. Reactions to optokinetic stimulation.
    V. Nystagmus in the dark.
   VI. Nystagmus and binocular perception.
  VII. Optical localization in patients with nystagmus.
 VIII. Latent nystagmus in one-eyed persons.
   IX. Gaze tonus and nystagmus.
    X. Causes of pathological pendular nystagmus.
   XI. Pendular nystagmus and latent nystagmus.
  XII. Causes of pathological jerking nystagmus, especially latent nystagmus

I. Refraction and visual acuity.

In the discussion of the individual cases we have never drawn special attention to the refraction. There is indeed not much to be said on this point, although the refraction does give rise to one or two remarks. Table I shows the refraction as found in the 6 different groups of the preceding chapter:

TABLE I

	1st. gr.	2nd. gr.	3rd gr.	4th. gr.	5th. gr.	6th. gr.	together
Emmetropia	4	2	4	15	3	3	31
Hypermetropia	6	5	19	8	5	4	47
Myopia	6	2	7	3	5	0	23
Isometropia	8	3	14	10	2		37
Anisometropia	0	1	1	3	4		9

Refraction of patients with nystagmus

Under the heading 'emmetropia' are included all eyes the refraction of which did not exceed 1 dioptre hypermetropia or myopia. If we had limited this group to pure emmetropia, the number of emmetropic eyes would have been no more than 9, belonging to only 5 individuals. The refraction of 101 eyes was noted; this number is too small to permit definite conclusions, but we gained the impression that there were relatively too few emmetropes among them. Hemmes (1924) was also struck by the fact that emmetropia occurred so seldom among his patients with hereditary nystagmus. Emmetropization (Straub, 1909) can hardly be regarded otherwise than as the consequence of a constant good accommodative adjustment of the eyes. In nystagmus patients with their disturbed optomotor reflexes and mobile eyes there will be difficulty in maintaining the correct accommodative adjustment and the emmetropization will thus be impeded. Keiner (1951) gave a similar explanation for the persistence of hypermetropia in many cases of strabismus convergens, here again the insufficient development of the optomotor reflexes is believed to impede the process of emmetropization.

In the second place we are struck by the fact that in the 3rd group we find, with 4 emmetropes, 19 hypermetropes and 7 myopes, whereas in the 4th group with 15 emmetropes we have only 8 hypermetropes and 3 myopes. Is this higher proportion of emmetropes in gr. 4 due to a better emmetropization in the representatives of this group? This does not appear entirely impossible. Group 4 comprises the cases with unequal optical gaze tonus of the 2 eyes. In many cases this inequality consisted in a stronger or more balanced development of the optomotor reflexes for the better eye. It is possible that this may have influenced the refraction. If this is so, we must also assume that the better development of these reflexes and the consequently better fixation with the one eye had also influenced the ultimate refraction of the other eye; probably owing to a better and more stable binocular accommodative adjustment.

In the third place we noticed that there were very few anisometropes among our patients — at any rate fewer than we had expected. The most pronounced anisometropia was seen in 4 patients of the 5th group; these had, in addition to an asymmetric optical gaze tonus, also an asymmetric non-optical gaze tonus. In Gr. 5 we found anisometropia in 4 of the 6 patients, while in the 40 patients of all the other groups together it occurred only 5 times. We feel, however, that these figures are too small to justify speculative deductions.

More closely and directly related to the development of the optomotor reflexes is the visual acuity. This is really a form of optical localization. Optical localization has as its physiological correlate the optomotor impulses which emanate from each part of the retina or from each retinal element in response to stimulation. In this way there is formed a tension pattern that maintains the optical gaze tonus. If an impulse from a peripheral part of the retina acquires some degree of predominance, in consequence of its own quality or via associative paths, so that other impulses are inhibited, we then speak of an adjusting impulse and an adjusting movement ensues. In the foveal region the optomotor impulses of different qualities (potential adjusting impulses) which arise from each of the cones constitute the physiological correlate of the visual acuity. Since visual acuity is a monocular function we can assume that it has developed from the monocular optomotor impulses. On these grounds we are of the opinion that a bad visual acuity with an otherwise intact retina is due to a faulty gradation of optomotor impulses from the cones of the fovea.

The causes of this faulty gradation may be of very different kinds and may lie anywhere in the reflex path that must be traversed by the stimuli before they can give rise to a cortical motor impulse. If no peripheral cause can be found for the poor visual acuity, a more central, cerebral cause becomes highly probable. In such cases a poor visual acuity affords support for the assumption of an insufficient development of the monocular optomotor reflexes.

As a rule we find with normal eyes after careful correction

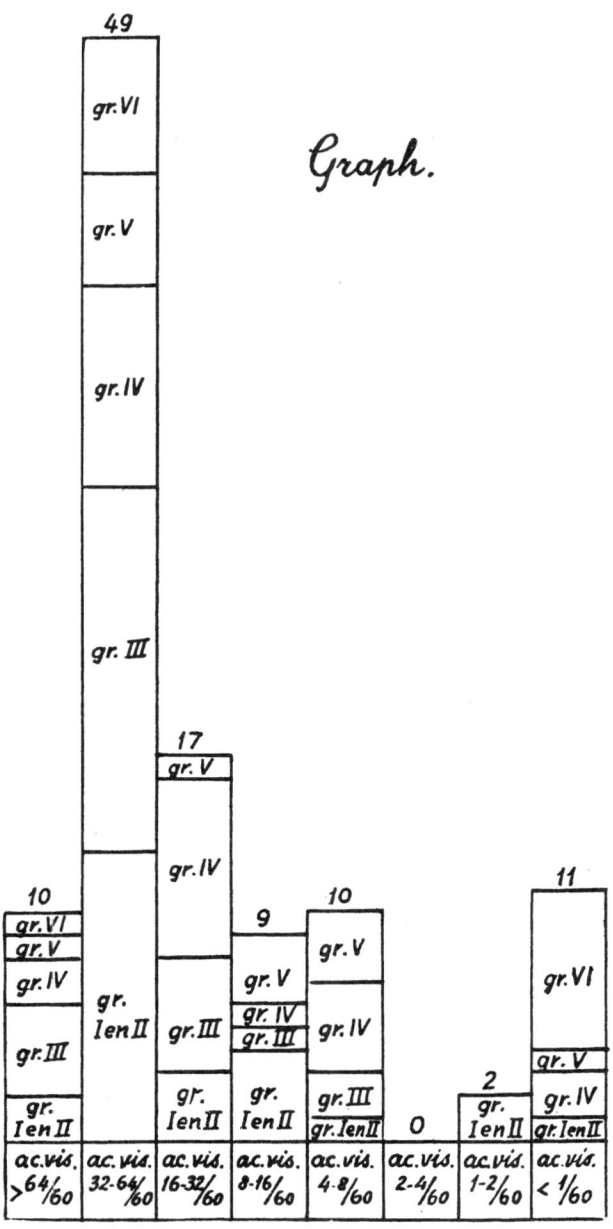

Graph.

Graph of visual acuity findings for 108 eyes of 54 patients with nystagmus (Gr. I and II pendular nystagmus; Gr. III, IV, V and VI latent nystagmus).

of errors of refraction a visual acuity of $^5/_4$. In our patients, such a visual acuity was found in only 10 of 108 eyes. These eyes belonged to 8 patients; there were thus only 2 with a visual acuity greater than 1 in both eyes. Among the 8 patients with very good visual acuity there was only one with pendular nystagmus (Case 2), and in this case the eyes were sometimes practically at rest. In 3 cases (Cases 39, 46 and 49) the eyes were at rest not only in binocular vision but also in monocular vision with the good eye; this makes the good visual acuity understandable. In 4 patients, however, (Cases 19, 21, 24 and 32) the eyes were at rest in binocular vision but not in monocular vision, and yet the visual acuity of the better eye was $^5/_4$. This shows that if the nystagmus movements are not too marked a good visual acuity is quite possible. Thus, although as a general rule we find a somewhat reduced visual acuity in patients with nystagmus, it is not by any means certain that the nystagmus movements are the cause and it is necessary to consider carefully whether the reduced visual acuity and the nystagmus are not both consequences of one and the same disturbance.

In latent nystagmus especially one would expect that the visual acuity in monocular vision would be considerably worse, on account of the nystagmus movements, than with binocular vision. The visual acuity was in some cases slightly higher binocularly than monocularly, but not so very often, while there was only one case of this kind in which a visual acuity of 1 was reached binocularly. This difference between monocular and binocular visual acuity is not done full justice to in the cases reported here, as the monocular acuity was often tested not with the other eye occluded but with a strongly positive lens before that eye, so that contours were still visible and the nystagmus movements of the fixating eye were thus inhibited. Our object was in the first place to ascertain the visual acuity of each eye without interference from nystagmic movements. Only 30 patients achieved a visual acuity of not less than 1 with binocular vision; 24 remained below this level. The fact that the visual acuity was on the whole rather low, even where no demonstrable ocular anomalies were present, is in our opinion due chiefly to a defective development of the monocular optomotor reflexes, especially from the central parts of the retina.

Also worthy of attention is the equality or inequality of the visual acuity of the 2 eyes. Of the 13 patients with pendular nystagmus, 6 showed a difference of some significance. With a pendular nystagmus one is inclined to expect a symmetrical tonic innervation, thus a symmetrical development of the optical reflexes and thus also an equal visual acuity. This, however, is not necessarily the case, if one eye has particular anomalies or is little or not used. The tonic innervation originating from the worse or non used (amblyopic) eye may then be weaker; this is sometimes demonstrated by larger excursions of the pendular nystagmus and by a weaker reaction to optokinetic stimulation in monocular vision with the worse eye (Cases 6, 8 and 9). A weaker reaction to optokinetic stimulation was found in several instances for the amblyopic eye of young squinting children, while at this early age it was also frequently observed that a contour movement in the temporal direction for the amblyopic eye produced a considerably weaker reaction than a contour movement in the nasal direction.

Considering now the 6 patients with unequal visual acuity we note that Cases 5, 6, 7, 8 and 9 had a convergent squint. Case 8 had moreover a cataracta punctata of the worse left eye, while Case 13 had very severe ocular anomalies: congenital aniridia and cataracta calcarea of the right eye. Only Case 4 showed no ocular anomalies that could account for the unequal visual acuity of the 2 eyes. A few words more about Case 9: this 3-year-old boy had strabismus convergens with amblyopia of the right eye. This right eye showed pendular nystagmus. After occlusion of the good eye for 2 periods of 4 weeks the amblyopia and pendular nystagmus of the right eye had disappeared, but then the left eye had become slightly amblyopic with pendular nystagmus. Now both eyes have a visual acuity of $^3/_4$ plus and the nystagmus has disappeared. The correlation between amblyopia and nystagmus is here unmistakable.

Of the 15 patients with latent nystagmus and practically symmetrical tonic gaze innervation, 11 had a practically equal visual acuity in both eyes and only 4 an inequality. Of these 4 patients 3 (Cases 18, 21 and 25) had strabismus. Only for the unequal visual acuity in Case 24 was it impossible to find a cause. As most of the patients in this group squinted, we cannot attach more value to this observation than to say that in cases of strabismus there is more chance of one eye being entirely neglected.

While in the group of patients with latent nystagmus and symmetrical tonic innervation only 26.7 % had unequal visual acuity, a quite different state of affairs was found in the patients with an asymmetric gaze tonus. Of the 20 patients, 15 (75 %) had unequal visual acuity. The question arises of whether the unequal visual acuity is the cause or the effect of the asymmetric gaze tonus, or whether both these disturbances are due to a common cause. In 8 cases there was every reason to assume that ocular anomalies were the cause of the unequal visual acuity and of the asymmetric gaze tonus. These were as follows: Case 31 with aphakia and R. microphthalmus accompanied by strabismus; Case 32 with cataracta zonularis accompanied by strabismus; Case 33 with slight atrophy of the optic nerve; Case 34 with congenital cataract of the R.E. accompanied by strabismus; Case 41 with cataracta punctata accompanied by strabismus; Case 44 with albinism; Case 46 with sequelae of iris prolapse accompanied by strabismus and Case 48 with coloboma iridis accompanied by strabismus. Of the remaining 7 cases with unequal visual acuity, 6 had strabismus (Cases 29, 30, 38, 40, 42 and 45). Only in Case 39 were there no anomalies which could be held responsible for the unequal visual acuity. We have seen, however, that strabismus as such does not by any means always give rise to an asymmetric gaze tonus. We therefore consider it very probable that in these 7 cases both the unequal visual acuity and the asymmetric gaze tonus were consequences of a more central disturbance in the development of the optomotor reflexes.

From our data on the refraction and visual acuity in patients with pendular nystagmus and latent nystagmus we come to the following conclusions:

(1) In patients with nystagmus the maintenance of the accommodative adjustment is rendered difficult by the constant movement of the eyes, so that the normal emmetropization is impeded and these patients will therefore show more errors of refraction.

(2) It is probable that a better and more balanced development of the optomotor reflexes of the eyes, or at any rate of the master eye, will promote emmetropization.

(3) The visual acuity of the great majority of nystagmus patients lies below the normal level. In many cases it is not possible to ascribe this reduced visual acuity to the nystagmic movements.

(4) An insufficient visual acuity without discernible cause is suggestive of an insufficient development of the monocular optomotor reflexes, especially from the central parts of the retina.

(5) With symmetrical development of the optomotor reflexes

from both eyes the chance of equal visual acuity of the 2 eyes is greater.

(6) An asymmetric development of the optomotor reflexes from the 2 eyes is often accompanied by unequal visual acuity of the 2 eyes.

II. Strabismus and alternating hyperphoria.

When we consider the fact that of our 55 patients with nystagmus 28 had a convergent and 9 a divergent squint, it seems an obvious conclusion that some relationship exists between nystagmus and strabismus. This relationship points to a related aetiology and since it is our conviction (Keiner 1951) that strabismus is the consequence of a disturbance in the development of the optomotor reflexes, it appears probable that nystagmus must also be due to a disturbed development of the optomotor reflexes. Further, we have frequently held a disturbance in the development of the monocular optomotor reflexes to be responsible for strabismus, and thus the question arises of whether this same disturbance can also be the cause of nystagmus, or whether it is necessary to seek some other explanation for the concomitance of strabismus and nystagmus.

Of the 13 patients with pendular nystagmus, 7 had strabismus, 5 had no strabismus and in one child we were unable to judge with certainty. This shows clearly enough that strabismus cannot be the cause of nystagmus any more than nystagmus is the cause of strabismus; there were too many nystagmus patients without strabismus.

Among the 13 patients with pendular nystagmus there was only one (Case 9) who appeared to have a disturbance of the development of the monocular optomotor reflexes exclusively. The normal optokinetic nystagmus and absence of a jerky terminal-position nystagmus indicated a satisfactory development of the conjugated optical fixation reflexes. The perfect parallelism between nystagmus and amblyopia in this child pointed to a disturbance in the development of the monocular optomotor reflexes, more especially those from the central parts of the retina, as the cause of the nystagmus.

Among the 13 patients with pendular nystagmus there were 2 (Cases 2 and 3) who showed evidence of a disturbance exclusively in the development of the conjugated optical fixation reflexes. The absence of strabismus and the good visual acuity of both eyes made a disturbance of development of the monocular optomotor reflexes very unlikely. The abnormal reaction to opto-kinetic stimulation and the presence of a definite terminal-position nystagmus indicated a not unimportant disturbance in the development of the conjugated optical fixation reflexes.

In the other 10 patients with pendular nystagmus both the monocular optomotor reflexes and the conjugated optical fixation reflexes were insufficiently developed. We come thus to the conclusion that a pendular nystagmus can develop as a result of a disturbance of development affecting exclusively the monocular optomotor reflexes or one affecting exclusively the conjugated optical fixation reflexes, but that in the majority of cases it will be caused by a combined disturbance of the 2 reflex systems.

This combined disturbance need not cause us any surprise if we consider the fact that the impulses in this reflex arc, at any rate as regards its afferent portion, follow a common path. Furthermore, the prompt occurrence of a fixation reflex in response to small displacements of the image on the retina will be greatly promoted by a sharp differentiation of the opto-motor stimuli from each of the retinal elements, to the development of which the monocular optomotor reflexes make an important contribution.

Considering now the 42 patients with latent nystagmus (including 7 one-eyed patients), we find in 22 of them a convergent squint, in 8 a divergent squint and in 10 no squint, while in 2 patients who had an artificial eye the presence of strabismus could not be ascertained. Among the patients without strabismus there was one (Case 53) with a suggestion of alternating hyper-phoria. We do not feel justified in concluding from these data that 55 % of individuals with latent nystagmus have strabismus convergens and 20 % have strabismus divergens. In the first place the number of patients is too small and in the second place the patients who consult an oculist always constitute a more or less selected group; few patients come to the oculist on account of their latent nystagmus; most of them come be-cause of a squint or other ocular affections. We do, however, feel justified in saying, but without giving definite percentages, that latent nystagmus is very frequently accompanied by strabismus and that this is usually a strabismus convergens. Nevertheless, the proportion of convergent to divergent strabismus is lower than in squinting patients without nystagmus. We can thus say that a correlation exists between the occurrence of strabismus and of latent nystagmus, but for the reason given above we cannot yet say that there is a special correlation between convergent strabismus and latent nystagmus. The predominance of conjugated fixation tonus in the nasal direction is thus not correlated with the predominance of monocular adduction tonus.

Perhaps more important is the fact that alternating hyperphoria was often observed in cases of latent nystagmus. Of our 42 patients with latent nystagmus 7 (Cases 17, 19, 20, 30, 43, 45 and 46) had a definite alternating hyperphoria, while in 5 others (Cases 27, 29, 42, 53 and 55) there was sufficient evidence for assuming that alternating hyperphoria was present here also, so that this affection was found in 28.6 % of the cases. Verhage (1941, 1942) also noticed that about $^1/_3$ of his patients with latent nystagmus showed alternating hyperphoria. The correlation between latent nystagmus and alternating hyperphoria has been brought to our knowledge chiefly by the investigation of Crone (1952). In 113 cases of alternating hyperphoria he found latent nystagmus 80 times, i.e. in about 71 % of the cases. Of these 80 patients, 67 had strabismus convergens (about 84 %) and 4 strabismus divergens (5 %), while 9 had no strabismus (about 11 %). To our small series of 12 patients with alternating hyperphoria we can add 3 others who had a pendular nystagmus as well. Of these 15 patients 11 had strabismus convergens (about 73 %), 2 strabismus divergens (about 13 %) and 2 no strabismus (about 13 %). One gets the impression that the cases of latent nystagmus in which there is also an alternating hyperphoria stand a much greater chance of having a convergent squint as well than do the cases of latent nystagmus without alternating hyperphoria.

The explanation given by Crone for the syndrome of alternating hyperphoria, to which latent nystagmus and convergent strabismus also belong, was the following: In these patients there is a defective development of the monocular optomotor reflexes which originate from the lower nasal quadrants of the retinae. As a consequence of this there is a deficiency of sursumduction, abduction and exorotation for each eye, with a resulting predominance of deorsumduction, adduction and endorotation. This, however, is only the case so long as the eye in question is open. If one eye is closed it will turn upwards in exorotation (alternating hyperphoria), partly under the influence of corrective conjugated impulses from the fixating eye. The open eye will also overcome the tendency to deorsumduction by a conjugated impulse to sursumversion. This impulse is supplied chiefly from the temporal lower quadrant. Coordinated with this, however, there occurs at the same time a conjugated innervation as a result of which the open eye moves nasalwards. To maintain or restore the intentional direction of gaze a conjugated innervation impulse is then given whereby the open eye is directed temporalwards. Thus there develops a conflict situation which manifests itself as latent nystagmus.

But apart from this conflict situation that arises in monocular vision, we must not forget that when both eyes are open stimuli also originate chiefly from the temporal lower quadrant to keep the gaze at the desired level. This means that from the left eye there comes a stimulus to dextroversion and from the right eye a stimulus to sinistroversion. In binocular vision this will do no harm; on the contrary it will raise the tonic innervation whereby the eyes are kept at rest. But if one eye is closed, this tonic innervation from the open eye has a disastrous effect.

Only in those cases in which not only the reflexes from the nasal lower quadrants but also those from both temporal quadrants have failed to develop properly can a divergent strabismus occur. The chance of latent nystagmus is then, however, much smaller because the predominance of the temporal lower quadrant over the nasal lower quadrant is then less great.

This explanation of the origin of latent nystagmus is so attractive that one would gladly apply it to all cases of latent nystagmus. We have indeed observed patients with latent nystagmus in whom an alternating hyperphoria had at first escaped recognition. But in view of the fact that alternating hyperphoria was found only 12 times in 42 unselected cases of latent nystagmus, it would not be justifiable to conclude that alternating hyperphoria is always the cause of latent nystagmus (even though it is possible that we may have missed some cases of alternating hyperphoria), unless it could be shown convincingly that the phenomenon of alternating hyperphoria can disappear while that of latent nystagmus remains. It is true that alternating hyperphoria can disappear as a result of the gradual development of a hyperfunction of the two inferior oblique muscles; in this way the deficiency of sursumduction and exorotation is abolished and the strabismus convergens persists. Nevertheless, with the development of hyperfunction of the inferior oblique the above-mentioned conflict situation, the primary factor in the latent nystagmus, is abolished, although the tonic innervation that results from it might be maintained. But a bilateral hyperfunction of the inferior oblique muscles is characterized by a strabismus sursoadductorius, and by no means all of our patients with latent nystagmus without alternating hyperphoria had this.

The 30 cases of latent nystagmus without alternating hyperphoria therefore require a somewhat different explanation. Of these 30 patients 13 had strabismus convergens, 6 strabismus divergens and 9 no strabismus, while each of the remaining 2 had an artificial eye. Now that the patients with alternating hyperphoria have been left out we have relatively more patients with strabismus divergens. [1]) Thus latent nystagmus and strabismus frequently go together, although we are not justified in saying that there is a preference for strabismus convergens. This means that latent nystagmus is very frequently seen in company with a disturbance of the development of the monocular optomotor reflexes. But such a disturbance is not always present. In Cases 14, 35 and 49 there was not the slightest evidence of a disturbance in the monocular optomotor reflexes, while in 7 other patients (Cases 24, 26, 28, 50, 51, 53 and 55) the monocular optomotor reflexes could not have been more than very slightly disturbed, although they showed a latent nystagmus that in several cases was quite strong. This makes it difficult to believe that latent nystagmus can be directly due to a disturbance in the monocular optomotor reflexes.

The jerking nystagmus of latent nystagmus is a binocular movement; the eye is incapable of fixating an object steadily. There is thus obviously a defective optical fixation mechanism. This view is supported by the reactions of these patients to optokinetic stimulation and by the presence of a pathological terminal-position nystagmus. We can safely say that all patients with latent nystagmus have a disturbed development of their conjugated optical fixation reflexes. But how are we now to regard the connection between this disturbance and that of the monocular optomotor reflexes? We have already pointed out that the disturbance in these monocular reflexes cannot be the direct cause of the disturbance in the conjugated reflexes. This makes it highly probable that both the developmental disturb-

---

[1]) Lagleyze (1913) found 3067 with strabismus convergens and 666 with strabismus divergens among 3791 squinting patients.

ance of the monocular optomotor reflexes and that of the conjugated optical fixation reflexes are derived from a very general disturbance in the development of the optical reflex tracts. Keiner (1951) suggested, for instance, a delayed myelination. We still have to consider whether a disturbance of the monocular optomotor reflexes may possibly interfere with the normal development of the conjugated optomotor reflexes.

This last idea is in accordance with the view of van der Hoeve (1917). He assumed primarily a labile equilibrium of the tonic gaze innervation. As a result of this lability, he believed that a normal predominance of reflexes to sinistroversion from the R. eye and to dextroversion from the L. eye became manifest and revealed itself as latent nystagmus. This presumptive predominance, however, has never been proved to exist in normal persons. We are therefore more inclined to regard such a predominance also as pathological, so that in the majority of cases both the monocular and the conjugated reflexes are primarily disturbed. This does not detract from the fact that Keiner (1951) came to the conclusion that the monocular reflexes from the temporal half of the retina develop earlier than those from the nasal half, also in normal children, and that if the development is retarded this often leads to strabismus convergens. It is also conceivable that the conjugated reflexes evoked by displacement of contours over the retina in a temporal direction develop earlier than those evoked by displacement of contours in a nasal direction and that in this way a latent nystagmus might arise with a delayed development of the conjugated reflexes. A good and prompt action of the fixation mechanism demands sharply differentiated optomotor reflexes from the different retinal elements, to which differentiation the monocular optomotor reflexes will undoubtedly also contribute. A disturbance in the development of the monocular optomotor reflexes might thus promote delay in the development of the conjugated optomotor reflexes, but it still cannot be the cause of latent nystagmus because there are too many cases of severe latent nystagmus in which the monocular optomotor reflexes show little or no disturbance.

Summing up we feel justified in saying that the following points have emerged from our study of the concomitance of strabismus and nystagmus:

(1) Strabismus is a frequent phenomenon in cases of nystagmus.

(2) Pendular nystagmus may be caused by a disturbance in the development of the monocular optomotor reflexes and also by a disturbance in the development of the conjugated optomotor reflexes, but most frequently it is due to a disturbance in the development of both reflex systems.

(3) In patients with latent nystagmus the excess of frequency of strabismus convergens over that of strabismus divergens is less great than that habitually found. There is thus no certain correlation between latent nystagmus and strabismus convergens.

(4) Latent nystagmus is found in a large majority (about 71 %) of patients with alternating hyperphoria. Conversely, we found alternating hyperphoria in 28.6 % of our patients with latent nystagmus.

(5) According to Crone (1952) the latent nystagmus in cases of alternating hyperphoria can be ascribed to a defective development of the optomotor reflexes from the two lower nasal quadrants of the retinae.

(6) In the majority of cases we must assume that the disturbance in the development of the monocular optomotor reflexes (strabismus) and the disturbance in the development of the conjugated optomotor reflexes (latent nystagmus) occur together and probably have a common cause.

(7) The possibility is worth considering that in normal children also the conjugated fixation reflexes to temporalward displacement of the retinal image may develop earlier than those to nasalward displacement. A delay in the development of these reflexes might be promoted by a disturbance in the monocular optomotor reflexes.

III. Nystagmus on looking sideways.

In order to understand the nature of nystagmus, it is necessary to observe it in the different directions of gaze. Although we have not yet progressed so far that we are able to give a satisfactory explanation of all the changes observed and to make use of these for a detailed diagnosis, yet it is worth while at this stage to describe and classify these changes and to explain them in an acceptable manner as far as possible.

As the most important changes were seen when the patients looked to the right and to the left, we have confined ourselves to a description of the nystagmus in sideways directions of gaze. In the case of pendular nystagmus we can expect only a change in the form or character, but in cases of latent nystagmus it is often only in sideways gaze that we see a nystagmus appear when the patient has both eyes open. In this latter case we may speak of a terminal-position nystagmus.

A typical terminal-position nystagmus is a jerking nystagmus with the fast phase in the direction of gaze. It always indicates a relative deficiency of fixation tonus. This deficiency may be due to an insufficient development of the optical fixation reflexes, but it may also be due to excessive demands being made on the fixation tonus. These excessive demands may be due to prolonged maximally lateral direction of gaze (physiological terminal-position nystagmus which occurs in about 60 % of all human beings: Offergeld 1893; Duke Elder 1949) and further to paresis of the ocular muscles, to gaze paresis or to an insufficient tonic innervation of either optical or non-optical character.

A pendular nystagmus on looking sideways is practically only seen in cases in which it was already present on looking straight ahead. In one patient (Case 22) we did indeed observe a pendular nystagmus exclusively in the adduction position with monocular vision. On looking straight ahead with the R.E. this patient showed a jerking nystagmus to the right; on looking to the left this changed into a pendular nystagmus. Probably the adduction position in monocular vision with the R.E. was the direction of the physiological resting position (neutral position) and the tonic innervation in this position was insufficient to keep the eyes at rest.

Returning now to the causes of jerking nystagmus in sideways direction of gaze, we must point out that ocular muscle pareses and gaze pareses were hardly seen at all among our 55 patients with nystagmus. Case 19 had a limitation of adduction of the R.E. and Case 42 had the same for the L.E., both as a consequence of a tenotomy of the internal rectus muscle of the eye in question. Case 29 showed a slight limitation of abduction of the squinting left eye and Case 31 had an abduction paresis of the L.E.

An insufficiency of tonic innervation can be either optical or non-optical in nature. The latter might be the case in affections of, for instance, the vestibular reflex pathways (sclerosis multiplex). In the majority of our cases, however, the nature of the nystagmus and the reactions to optokinetic stimulation showed clearly that the disturbance must be sought in the optical reflex pathways.

If the insufficient tonic innervation of optical nature concerns more especially the light tonus, then in practically all cases both the tonic innervation to right turning and the tonic innervation to left turning are insufficient in all directions of gaze so that a pendular nystagmus results. Even if the optical fixation reflexes have developed satisfactorily under these conditions, a pendular nystagmus may be present in all directions of gaze. Examples of this are seen in Cases 5, 6 and 9.

The possibilities that may arise in sideways gaze are thus as follows: (1) no nystagmus: seen in cases of normal and very slightly disturbed development of the optomotor reflexes; (2) weak jerking nystagmus in the direction of gaze with a slightly more severe disturbance of the optical fixation reflexes; (3) lively jerking nystagmus with more severe disturbance of the optical fixation reflexes in the direction of gaze; (4) pendular nystagmus with a disturbance in the tonic innervation of the eyes (defective light tonus).

In patients with pendular nystagmus there is as a rule no predominance of the gaze tonus in a given direction, for which reason the nystagmus on looking to the R. and to the L. shows complete symmetry. If the optical fixation reflexes are not disturbed, the pendular nystagmus persists also with sideways direction of gaze; if the optical fixation reflexes are disturbed, a jerking nystagmus appears. Only in a few cases (Cases 10 and 11) was this jerking nystagmus not symmetrical to both sides. It was found, however, that in both these cases the physiological resting position did not exactly coincide with the straight ahead direction of gaze. In Case 10 the eyes were directed slightly to the right in this resting position and in Case 11 slightly to the left. This can be called an eccentric neutral position (Anderson, 1953). Less easily explained is the asymmetry for sideways direction of gaze in Case 13. This girl had a jerking nystagmus to R. on looking to the R. and a pendular nystagmus on looking to the L. She had very bad sight, the R.E. being practically blind. All the evidence pointed to a very poor development of her optomotor reflexes. It is probable that only the fixation tonus to the L. from the L.E. had developed somewhat better, so that on looking to the R. this fixation tonus plus the tension in the tissues caused the jerking nystagmus, while on looking to the L. the fixation tonus and the tension in the tissues more or less balanced each other so that the pendular nystagmus persisted.

With monocular vision one would also expect a symmetrical disturbance on looking to the R. and to the L. in cases of pendular nystagmus. In the great majority of cases this is so. In addition to the Cases 10, 11 and 13 mentioned above, exceptions to this rule were seen in Case 1 (R.E.), Case 2 (R.E.), Case 3 (L.E.) and Case 6 (both eyes). With all these eyes a stronger jerking nystagmus appeared on looking temporalwards than on looking nasalwards. This indicates that the temporally directed fixation tonus had lagged further behind in development than the nasally directed tonus, so that one might say that in these cases there was an inapparent latent nystagmus. Cases 1, 2, 3 and 6 thus represented transitional forms between pendular nystagmus and latent nystagmus.

If we now consider what may be expected in latent nystagmus — leaving the one-eyed patients out of account for the moment — we find that this depends on two factors: (1) the degree of disturbance in the development of the optomotor reflexes and (2) the presence of symmetry or asymmetry in this disturbance of the optomotor reflexes for the two eyes. On the grounds of the phenomena observed in our patients we have subdivided the cases of latent nystagmus into 3 groups: Group III comprises those in which the innervation tonus of the eyes had developed more or less symmetrically; Group IV those in which the non-optical gaze tonus had developed symmetrically but the optical gaze tonus had not and Group V those in which both the optical and the non-optical gaze tonus had developed asymmetrically.

Only with a very slight degree of disturbance of the development of the optomotor reflexes will the eyes remain at rest on looking sideways with both eyes. We found this in Cases 14, 15, 16 and 39.

With a symmetrical development of the optomotor reflexes we may expect, also in cases of latent nystagmus, that no difference will be seen in looking with both eyes to the right or to the left. It is therefore not surprising that of the 15 patients in group III only 3 showed an asymmetry in this respect, whereas in the groups IV and V the proportion was 9 of the 20 patients. The question that does arise is rather the following: what was the reason for the 3 cases of asymmetry in gr. III and why were there not more cases of asymmetry in gr. IV and gr. V?

These 3 cases in gr. III were Cases 21, 26 and 28. The gaze tonus was indeed not perfectly symmetrical in Case 21 (this patient kept the head turned somewhat to the left) or in Case 26; in both the physiological resting position was slightly to the right. We are unable to offer any explanation for the asymmetry in Case 28. In spite of this asymmetry we have placed these 3 cases in group III because the optokinetic nystagmus to the R. and to the L. was completely symmetrical.

With regard to the question of why groups IV and V contained such a relatively large number of patients with a symmetrical behaviour in binocular sideways gaze, the following seems very probable: For each eye the gaze tonus in the horizontal plane is the resultant of a component to right turning and a component to left turning. It seems quite possible that in the preferred eye the optomotor reflexes will be better developed, but that the difference between the 2 components is the same as that in the less favoured eye. The 2 resultants then balance each other, although the components are not the same for each eye. Cases 36, 38 and 40 are examples of this. It is also possible that the more favoured eye has developed normally and is strongly dominant in binocular vision (Case 39).

In patients with latent nystagmus, no symmetry is to be expected in looking sideways with only one eye open; the jerking nystagmus will always be stronger on looking in the temporal direction because the fixation tonus is practically always more disturbed in this direction. We found only the followng exceptions: Case 39 (R.E.), Case 44 (L.E.), Case 45 (L.E.), Case 46 (L.E.) and Case 48 (R.E.). Case 39, however, did not constitute an exception because there was no nystagmus in monocular vision with the good R.E., showing that the optomotor reflexes of this eye had developed practically normally. In Cases 44, 45, 46 and 48, the excess of nasalward optical gaze tonus was apparently counterbalanced completely by the temporalward non-optical gaze tonus.

A quite different state of affairs is encountered with the nystagmus in monocular nasally-directed gaze. Here we can expect almost anything: (1) a jerking nystagmus nasalward; (2) total disappearance of the nystagmus; (3) jerking nystagmus temporalward; (4) pendular nystagmus. A nasalward jerking nystagmus will be found in cases where the nasalward fixation tonus has also developed insufficiently and the difference in fixation tonus temporalwards and nasalwards is not great. Total disappearance of the nystagmus indicates a good development of the fixation tonus nasalwards, while the difference between this and the fixation tonus temporalwards is not so very large. A temporalward jerking nystagmus points to a very large difference between the nasalward and the temporalward fixation tonus and is to be expected particularly in cases where the nasalward optical fixation tonus is still further strengthened by an excess of non-optical fixation tonus in the same direction (Cases 44, 45, 46 and 48). A pendular nystagmus is a sign of very low tonic innervation and also indicates that the difference between the nasalward and the temporalward fixation tonus is not great; it is further particularly to be expected where this small difference is compensated by a non-optical fixation tonus in the opposite direction (Case 50).

A difference between the R. and L. eye in temporalward and nasalward gaze is not likely in Group III on account of the symmetrical development of the optomotor reflexes. There are only a few exceptions, including once more Cases 26 and 28. We are of the opinion that in these exceptional cases the development of the optomotor reflexes had not really progressed quite symmetrically.

A difference of this kind between the R. and L. eye becomes much more likely in the groups IV and V. In group IV it is still conceivable that the difference of fixation tonus for right and left turning is the same for both eyes but that the components which give this difference are not identical. In such a case the possibility of a symmetrical behaviour of the 2 eyes in sideways gaze is not excluded. In group V this is no longer conceivable. The asymmetric non-optical gaze tonus (manifested by jerking nystagmus in the dark) will weaken the tendency to nystagmus in monocular vision with the one eye and strengthen it in monocular vision with the other eye. This will inevitably lead to a different behaviour of the 2 eyes in monocular sideways gaze with each eye in turn. This phenomenon was seen in all patients of group V.

Group VI, the group of one-eyed patients, corresponds perfectly to group V. For the seeing eye there is in these patients, as in practically all patients with latent nystagmus, an excess of optical fixation tonus in the nasalward direction, but in addition to this an excess of non-optical gaze tonus in the temporalward direction. Thus the tendency to temporalward jerking nystagmus will become greater on looking temporalwards and will disappear on looking nasalwards, or may even become transformed into a nasalward jerking nystagmus if the asymmetry of the non-optical gaze tonus is very marked (Cases 49 and 54).

In the foregoing we have discussed the significance of the presence or absence and the nature of the nystagmus on looking sideways and we can now summarize our conclusions, both from theoretical considerations and from the results of examination of our patients, as follows:

(1) A jerking nystagmus on looking to the side is due to an insufficiency of the optical fixation reflexes. This insufficiency may be absolute or relative.

(2) A pendular nystagmus in looking to the side is a sign that the tonic gaze innervation in general is too low; this is not necessarily due in the first place to disturbance of the optical fixation reflexes.

(3) Symmetrical behaviour of the eyes in binocular gaze to the right and to the left is a probable but not a certain sign of a symmetrical development of the optomotor reflexes.

(4) Asymmetric behaviour of the eyes in binocular gaze to the right and to the left is proof of an asymmetric development of the optomotor reflexes.

(5) Identical behaviour in monocular gaze to nasal and to temporal is a probable but not a certain sign of equal development of the nasally and temporally directed optical fixation tonus.

(6) Non identical behaviour in monocular gaze to nasal and to temporal is proof of an unequal development of the nasally and temporally directed optical fixation tonus.

(7) Identical behaviour of the right and the left eye in monocular sideways gaze is a probable but not a certain sign of symmetrical development of the optomotor reflexes from the two eyes.

(8) Non-identical behaviour of the right and the left eye in monocular sideways gaze is proof of an unequal development of the optomotor reflexes in the two eyes.

(9) With an asymmetric development of the non-optical gaze tonus the possibility of identical behaviour of the right and left eye in monocular sideways gaze is excluded.

IV. **Reactions to optokinetic stimulation.**

In all our 55 patients with nystagmus the optokinetic nystagmus in the horizontal direction was examined both in binocular and in monocular vision.

The manner in which, in our opinion, the optokinetic nystagmus is produced has been explained in detail in the preliminary notes: the slow phase is due to a unilateral elevation of the fixation tonus under the influence of the optical fixation reflexes; the fast phase is the result of numerous subcortical reflexes to which also the light tonus and under certain conditions also the optical direction and adjusting reflexes contribute.

There is a certain parallelism between the investigation of the reactions to optokinetic stimulation and the investigation of the terminal-position nystagmus; both were examined binocularly and monocularly in two directions. In the optokinetic nystagmus examination, however, a well-marked jerking nystagmus is a sign of good functioning of the conjugated optical fixation reflexes, whereas in the examination of terminal-position nystagmus a well-marked jerking nystagmus was just a sign of very poor functioning of these fixation reflexes.

The different modes of reaction to optokinetic stimulation are as follows: (1) A more or less strong jerking nystagmus in the opposite direction to that in which the contours move; (2) no movement of the eyes; (3) jerking nystagmus in the direction of movement of the contours; (4) pendular nystagmus. In order to evaluate these reactions it is necessary to know whether nystagmus was already present without optokinetic stimulation and, if so, whether this was a pendular or a jerking nystagmus. Only then is it possible to judge whether one is dealing with a normal optokinetic reaction, with optokinetic insensitivity ('optische Drehstarre') or with an abnormal type, i.e. a true inverse type.

In our discussion we shall deal in turn with (A) the patients with pendular nystagmus, (B) those with latent nystagmus and (C) the one-eyed patients with latent nystagmus.

A. *Patients with pendular nystagmus.*

In cases of pendular nystagmus, which so often must be ascribed to ocular anomalies (maculae corneae, congenital cataract, fundus anomalies, albinism), a disturbance in the development of the optical fixation reflexes is very likely. Since the existence of optical fixation reflexes is unconditionally necessary for the production of optokinetic nystagmus, a disturbance of this reflex mechanism will weaken the optokinetic nystagmus and in severe cases will render the optokinetic stimuli totally ineffective ('optische Drehstarre') or will give rise to an inverse type. The results of our investigation did in fact confirm these expectations.

Of our 13 patients with pendular nystagmus only 3 (Cases 5, 6 and 9) showed a reasonably good optokinetic nystagmus; one patient (Case 1) had a weak but normal optokinetic nystagmus; 3 patients (Cases 2, 7 and 11) continued to display their pendular nystagmus alternating with a very weak normal jerking nystagmus; 4 patients (Cases 3, 4, 8 and 10) continued to display their pendular nystagmus and did not react at all to the optokinetic stimulation and one patient (Case 13) showed an inverse type. Case 12 showed with contour movement to the R. a very weak jerking nystagmus to L. and with contour movement to the L. a pendular nystagmus alternating with small jerks to L.

In connection with the examination of terminal-position nystagmus we have already pointed out that in Cases 5, 6 and 9 we were obliged to conclude that there was chiefly a disturbance in the development of the monocular optomotor reflexes (light tonus) and that the development of the conjugated fixation reflexes was only slightly defective. The same assumption must be made for Case 1, although here the disturbance of the conjugated optical fixation reflexes was shown both by the terminal-position nystagmus and by the optokinetic nystagmus to be somewhat more severe. In all the other patients the disturbance of the optical fixation reflexes was unmistakable and the anomaly was practically equal with contour movement to the R. and to the L. Only Case 12 failed to show this symmetrical reaction so that we are obliged to assume that he had a small excess of fixation tonus to the R., which would also account for the inverse type with contour movement to the L.

The optokinetic nystagmus with monocular vision is also important. The results obtained in this way are summarized in Table II. With monocular vision also, the optokinetic stimulus was ineffective in the majority of our patients with pendular nystagmus. We must point out, however, and this applies also to vision with both eyes, that the pendular nystagmus was often more lively during the optokinetic stimulation than in the absence of such stimulation. We surmise that the movement of the contours deprived the eyes of any point of support with the aid of which they could be kept still by adjusting movements.

TABLE II

Contour movement	practically normal jerking nystagmus	very weak jerking nystagmus	pendular nystagmus or eyes at rest	inverse type
temporalward	3	4	15	4
nasalward	6	3	16	1

Optokinetic reactions with monocular vision in patients with pendular nystagmus

Of the 31 observations which we have classified as no reaction to optokinetic stimulation, there were 9 in which a few jerks did appear from time to time; on 3 occasions these jerks were in the opposite direction to the contour movement and on 6 occasions in the same direction as the contour movement (inverse type), but in these cases the pendular nystagmus was strongly predominant.

A remarkable fact is that the inverse type occurred more frequently with movement temporalwards than with movement nasalwards. This suggests that with monocular vision the nasalward tonic innervation of the eyes, even with pendular nystagmus, still sometimes predominates over the temporalward tonic innervation. This too points to the existence of transitional forms between pendular and latent nystagmus. The numbers are of course too small to afford proof, but there are a few other phenomena which strengthen our conjecture.

In the first place we see that a practically normal jerking nystagmus occurred only 3 times with temporalward contour movement and 6 times with nasalward contour movement. This also suggests a predominance of the nasalward tonic innervation.

In the second place we found with 8 eyes that the optokinetic reaction was different for nasalward and temporalward movement of the contours. This difference was shown by the R.E. of Case 1, the L.E. of Case 2, the L.E. of Case 5, both eyes of Case 6, both eyes of Case 10 and the R.E. of Case 12. These differences pointed once more in Cases 1, 2, 5, 6 and 10 to a predominance of the nasalward tonic innervation. Only Case 12 was an exception; here it appeared that for the R.E. the temporalward tonic innervation predominated. Except in the last-mentioned case, thus, the difference in question was reminiscent of the condition in latent nystagmus.

It goes without saying that in such cases a difference in reaction between the two eyes will also be seen. A difference in reaction of the two eyes was further seen in Case 13: this patient's R.E. had very bad sight and therefore did not react at all to the optokinetic stimulation; her L.E., which was also far from normal, reacted in both directions with an inverse type.

B. *Patients with latent nystagmus.*

As the one-eyed patients will be discussed separately, we are concerned here with observations on 35 patients. The nystagmus in monocular vision shows unequivocally, irrespective of its origin, that there is then a difference between the temporalward and the nasalward tonic innervation. This, however, does not imply that with both eyes open the tonic innervations to the right and to the left must be different. This will only be the case if the tonic innervation of the two eyes has not developed symmetrically. One of the means of ascertaining this is the study of the optokinetic nystagmus, in which it is noted whether with binocular vision the contour movement to the right gives a reaction which is similar to (symmetrical with) that given by contour movement to the left.

In our 35 patients we saw 23 times a symmetrical reaction and 12 times an asymmetric reaction with both eyes open. A symmetrical reaction with binocular vision does not by any means always mean that the optical gaze tonus in the 2 eyes is symmetrically developed; it tells us only that the nasalward fixation tonus of the one eye combined with the temporalward fixation tonus of the other eye is equal to the temporalward fixation tonus of the one eye combined with the nasalward fixation tonus of the other eye. An asymmetric reaction to optokinetic stimulation with contour movement to the R. and to the L. is certain proof of an unequal gaze tonus in the 2 eyes. Nevertheless we have included in group III (patients with latent nystagmus and symmetrical gaze tonus) one patient (Case 22) who showed an asymmetric optokinetic nystagmus with binocular vision. In the discussion of the terminal-position nystagmus we have already pointed out that in our opinion this patient had a slight deviation (to the left) of the physiological resting position, while further the optical fixation reflexes to the right and to the left were symmetrically developed.

The nature of the reaction to optokinetic stimulation with binocular vision can also differ widely in patients with latent nystagmus. The findings in this connection are presented in Table III, in which the cases of group III and those of groups IV and V are kept separate.

TABLE III

Group and contour movement	practically normal jerking nystagmus	very weak jerking nystagmus	pendular nystagmus or eyes at rest	inverse type
Gr. III to R.	4	3	4	4
Gr. III to L.	4	4	4	3
Gr. IV and V to R.	10	6	3	1
Gr. IV and V to L.	11	3	3	3

Nature of optokinetic reactions with binocular vision in patients with latent nystagmus.

A practically normal jerking nystagmus to the R. and to the L. suggests that the tonic innervation in the temporalward direction has also developed more or less satisfactorily. This we saw in 4 patients from group III (Cases 14, 15, 18 and 20) and in 5 patients from groups IV and V (Cases 29, 31, 36, 42 and 46). There were, thus, in groups IV and V 11 patients whose optical fixation reflexes with binocular vision had only developed satisfactorily in one direction. An optokinetic insensitivity (eyes at rest or pendular nystagmus) and an inverse type with contour movement to the R. and to the L. suggest that the tonic innervation in the nasalward direction is also considerably deficient. This we saw in 7 patients from group III (Cases 21, 23, 24, 25,

26, 27 and 28) and in only one patient from groups IV and V (Case 32). Thus in 8 patients from groups IV and V the optical fixation reflexes in one direction were practically inactive with binocular vision, while in the other direction they were quite clearly demonstrable.

In order to obtain more exact information as to a possible symmetry or asymmetry in the development of the tonic innervation of the 2 eyes, and also as to the degree of development and amount of difference in the development of the tonic innervation, with special reference to the optical fixation tonus in the temporalward and nasalward directions, it is necessary to study the optokinetic nystagmus in monocular vision as well.

It seems obvious that with latent nystagmus the optokinetic reaction in monocular vision will practically always be different for temporalward and nasalward contour movement. An equal reaction under these conditions was in fact only rarely seen. We observed such a reaction with only 7 of the 70 eyes of our 35 patients: Case 28 for both eyes; Cases 31, 35 and 39 for the R.E. only and Cases 44 and 46 for the L.E. only. Let us now examine these exceptional cases more closely.

In Case 28 optokinetic stimulation gave rise to a pendular nystagmus on every occasion with monocular vision and to an inverse type with binocular vision. All this shows that not only the fixation tonus but the whole tonic innervation of the eyes was at a very low level. As a result of this, with nasalward contour movement the slight excess of fixation tonus in that direction soon became exhausted and there remained — as with temporalward movement of the contours — only a pendular nystagmus.

In Case 31 there was also only a very slight excess of the nasalward fixation tonus for the bad R.E. This eye showed practically no nystagmus in monocular vision. The tonic innervation of the eyes in general appeared to be at a somewhat higher level than in Case 28, so that the R.E. did not display a pendular nystagmus but remained still.

In Case 35 the weak optokinetic reaction, also in binocular vision, shows that the level of the fixation tonus was low. This must be held responsible for the fact that the slight excess of nasally-directed fixation tonus did not become manifest for the R.E. under optokinetic stimulation.

In Case 39 the optomotor reflexes from the R.E. had developed normally. This patient thus did not show any nystagmus in monocular vision with the R.E., so that a difference with temporalward and nasalward contour movement was not to be expected for this eye.

In Cases 44 and 46 the conditions were quite different. Here we found evidence suggesting an excess of non-optical gaze tonus to the L. In consequence of this the excess of nasalward optical fixation tonus was counterbalanced for the L.E. and reinforced for the R.E.; this also accounts for the large difference between the reactions of the R.E. and the L.E.

We can further ascertain how often the optokinetic reactions of the 2 eyes were perfectly equal, i.e. perfectly symmetrical, and how often this was not the case. We find that 11 of the 35 patients showed a practically symmetrical optokinetic reaction while the others did not. These 11 patients also had an exactly equal reaction for contour movement to the R. and to the L. with binocular vision; this symmetry and this equality are the chief reasons why these 11 cases have been placed in Group III (latent nystagmus with symmetrical gaze tonus).

In Table IV we give a survey of the various ways in which our patients reacted to optokinetic stimulation with monocular vision. In this connection it must be remembered that in the great majority of cases the starting point was a jerking nystagmus in the temporal direction. We have therefore paid more attention to the change of the nystagmus than to the nature of the nystagmus.

TABLE IV

Contour movement	jerking nyst. temp. enhanced	no change	jerking nyst. temp. weakened	no jerking nystagmus	jerking nyst. nasalward
temporalw.	7 inverse type	15 insensitive	20	17	11
nasalw.	33	25 insensitive	2 inverse type	7 inverse type	3 inverse type

Changes in the nature of the nystagmus on optokinetic stimulation with monocular vision in patients with latent nystagmus.

From this table we see that with temporalward contour movement 15 and with nasalward contour movement 25 eyes were practically insensitive to the optokinetic stimulation. We prefer not to say completely insensitive,as no nystagmograms were made, so that small changes may have been missed. A weakening or disappearance of an existing temporalward jerking nystagmus with contour movement in the temporal direction is to be expected. The 7 eyes in which the temporalward jerking nystagmus was further strengthened by temporalward contour movement (inverse type) and the 11 eyes in which the optokinetic nystagmus prevailed over the latent nystagmus must be further considered.

The table also shows that with a nasalward contour movement the optokinetic stimulation was able to reinforce the temporal jerking nystagmus of 33 eyes; 12 eyes showed an inverse type.

It thus appears that a temporalward contour movement can more easily weaken the existing nystagmus (48 eyes) than a nasalward contour movement can strengthen it (33 eyes). One might also say that the nasalward, relatively too high optical tonic innervation is more easely inhibited than further strengthened.

We shall now consider first the 11 cases in which with a temporalward contour movement the optokinetic nystagmus prevailed over the existing jerking nystagmus. These 11 eyes belonged to 9 different individuals.

Such a predominance of the optokinetic nystagmus appears to us to be conceivable under only 2 conditions: (1) if the excess of the nasalward optical fixation tonus is not very great and (2) if the excess of the nasalward optical fixation tonus is totally or partially compensated by a non-optical gaze tonus in opposite direction. This brings us to the question of whether it is possible to ascertain which of these 2 conditions exists in a given case. In most cases this is indeed possible. In the first place we must remember that a small difference in the optical fixation tonus for the one eye will have no influence in monocular vision with the other eye. An asymmetric non-optical gaze tonus, on the other hand, which has a compensating effect for the one eye, will greatly enhance the difference between nasalward and temporalward tonic innervation for the other eye, thus causing a markedly asymmetric optokinetic reaction of the 2 eyes. If, as in Cases 15 and 27, we find a normal optokinetic nystagmus with nasalward contour movement for both eyes, an asymmetric non-optical gaze tonus can be excluded.

In the second place, an asymmetric non-optical gaze tonus

can only have a compensating effect for one eye; it is to be expected that this will be the better, more used eye and not the worse, often squinting eye. Occlusion of the better eye will therefore give a much stronger jerking nystagmus than occlusion of the worse eye (Cases 44, 45, 46 and 48).

In the third place, a jerking nystagmus in the dark is proof that the non-optical gaze tonus is asymmetric.

On the basis of these considerations we may conclude that in Case 39 there was no difference between nasalward and temporalward fixation tonus for the R.E., that in Cases 15, 27, 35 and 36 this difference was only small and that in Cases 44, 45, 46 and 48 a compensatory non-optical gaze tonus on behalf of the better eye had developed.

Table IV shows 7 eyes with an accentuated jerking nystagmus to the temporal side with temporal movement of the contours. This must be regarded as an inverse type. It occurred in 6 individuals (Cases 14, 23, 26, 34, 37 and 48). An inverse type is only conceivable if the fixation tonus in the direction of the moving contours is absolutely very low and is not activated by the optokinetic stimuli but, on the contrary, is exhausted by them. A low fixation tonus in Cases 26 and 34 appeared very probable, as both these patients had had a pendular nystagmus in childhood.

The inverse type with contour movement in the nasalward direction is more remarkable, as in nystagmus latens one is obliged to assume that there is always an excess of fixation tonus in the nasalward direction. We found this in 12 eyes belonging to 7 patients (Cases 23, 26, 27, 28, 41, 45 and 48). In Cases 23, 26, 27 and 28 there was indeed a very low gaze tonus; Case 23 had nystagmus in the dark, Cases 26 and 28 still had pendular nystagmus in daylight and Case 27 had had pendular nystagmus as a child. These 4 patients showed the inverse type with nasalward contour movement, both when using the R.E. and when using the L.E.; this symmetry was one of the reasons for placing them in group III.

In Case 41 we observed an inverse type (decrease of the existing jerking nystagmus) with nasalward contour movement for both eyes, but not of equal strength for both eyes. The inverse type was more marked with the better R.E. In the dark the eyes were not completely at rest; here also we feel justified in assuming that the fixation tonus in all directions was very low. For Cases 45 and 48 a somewhat different explanation is needed.

In Case 45 the general gaze tonus was low, as shown by the pendular nystagmus. But this patient also had an asymmetric non-optical gaze tonus with a predominance to the L. Optokinetic nystagmus is brought about by an activation of the optical fixation tonus in the direction of the contour movement and an inhibition of the optical fixation tonus in the opposite direction. In nasalward contour movement for the L.E. of this patient there was but little temporalward fixation tonus to be inhibited, while the non optical gaze tonus remained unchanged. In this way a slight degree of exhaustion of the nasalward fixation tonus by the higher gaze tonus to the L. would very easily lead to a jerking nystagmus to the R. A nasalward contour movement for the R.E., on the other hand, would lead to an accentuated jerking nystagmus temporalwards, also as a consequence of the increased non-optical gaze tonus to the L. A similar explanation applies to Case 48; except that here the increased non-optical gaze tonus was directed to the R., so that the R.E. showed an inverse type with nasalward contour movement and the L.E. a strong temporalward jerking nystagmus. In Case 48 there was no jerking nystagmus in monocular vision with the R.E., but in monocular vision with the L.E. a jerking nystagmus was present.

Taking all these facts together we find in all instances that an inverse type is seen in patients with a very low fixation tonus; the occurrence of an inverse type may also be favoured by an asymmetric non-optical gaze tonus with a predominance in the opposite direction to that of the contour movement.

As for the inverse type, the insensitivity to optokinetic stimuli ('optische Drehstarre') is also a proof of deficiency of the optical fixation reflexes in the direction of the contour movement and the fixation tonus originating from these. On a total of 70 eyes, optokinetic insensitivity was observed on 40 occasions, in 28 eyes belonging to 19 different individuals. For the temporalward direction we found insensitivity in 15 observations on 11 patients and for the nasalward direction in 25 observations on 17 patients. The fact that in cases of latent nystagmus the jerking nystagmus with monocular vision is practically always temporalwards we consider to be due to a deficiency of fixation tonus temporalwards and in consequence of this a relative excess of fixation tonus nasalwards. If now the optical fixation reflexes in the temporalward direction are insufficient, it is not surprising that a contour movement in the temporalward direction gives no optokinetic reaction. What may seem more surprising is that the optokinetic reaction may also be absent when the contours move in the nasalward direction. But even though there may be a relative excess of fixation tonus in nasalward direction, the absolute value of this fixation tonus may still be rather low. We did in fact find that among the 25 eyes which showed no change in the jerking nystagmus when the contours moved nasalwards there were 12 which also showed no reaction to temporalward contour movement. In these eyes, thus, there was a low fixation tonus in both directions. We are then left with 13 eyes which did not react to nasalward contour movement and did react to temporalward contour movement. One could of course say that the jerking nystagmus of latent nystagmus is more easily weakened than strengthened by optokinetic stimulation, but if we look more closely into our cases we see that of these 13 eyes there were only 5 which belonged to patients — 3 in number — whom we considered to have a symmetrical gaze tonus and whom we had placed on this account in group III, while the other 8 eyes belonged to patients — 7 in number — in whom for various reasons it was concluded that an asymmetric optical gaze tonus existed and who had therefore been placed in group IV. Thus, if we find a patient with latent nystagmus who, in monocular vision, shows no reaction with contour movement nasalward but does show a reaction with contour movement temporalward, and further if this disturbance is confined to one eye, then there is a great likelihood that we are dealing with a patient who has an asymmetric optical gaze tonus.

This would appear to be a suitable point at which to deal in rather more detail with the asymmetric optical gaze tonus, because it is just the optokinetic reaction that shows up any asymmetry of this kind the most clearly.

When the eyes are at rest, and also when a pendular nystagmus is present, it may be said that the innervation to right turning and that to left turning just balance each other. This equilibrium is achieved on the one hand by an adjustment innervation and on the other hand by a tonic innervation. The adjustment innervation soon becomes unable to hold out against the tonic innervation, as is shown by the physiological terminal-position nystagmus that is seen in so many people in more or less prolonged sideways gaze. Thus, if no jerking nystagmus appears when the patient gazes straight ahead, we are justified in assuming that the tonic innervation also is such as practically to maintain the forward direction of gaze.

This tonic innervation again is made up of numerous components. For our purpose we shall divide these simply into optical and non-optical components. If the optical tonic innervation to right turning is not equal to the optical tonic innervation to left turning, then where this difference is of any appreciable size a jerking nystagmus cannot fail to appear unless this difference is compensated by non-optical reflexes. It is indeed possible for a slight asymmetry between the optical tonic innervations to right and left turning to exist without compensation by an asymmetric non-optical gaze tonus but, as already remarked, this difference cannot be a large one. In patients in whom we encountered such a state of affairs the eyes were often not absolutely at rest and small jerks were seen in the direction in which the optical tonic innervation was insufficient, while in other cases the head was kept turned in that direction.

It must also not be forgotten that the conjugated optical tonic innervation is the resultant of the following 4 components: (a) a tonic innervation to right turning originating from the right eye; (b) a tonic innervation to left turning from the right eye; (c) a tonic innervation to right turning from the left eye and (d) a tonic innervation to left turning from the left eye.

In latent nystagmus we have $b > a$ and $c > d$. If $b - a = c - d$, the eyes will remain at rest; but this does not necessarily mean that b is equal to c or a equal to d.

If the optical fixation reflexes and the associated fixation tonus have almost or entirely failed to develop, we must expect a pendular nystagmus without optokinetic reaction. Our impression is that the optical fixation reflexes in the nasalward direction develop first — a point to which we shall return later. If the optical fixation reflexes in the temporalward direction fail to catch up with those in the nasalward direction and if the development occurs symmetrically, we shall get a latent nystagmus with symmetrical gaze tonus (the cases of group III). Now it is possible that the optical fixation reflexes originating from the better or more used eye may reach a higher lever of development than those from the worse and often squinting eye. This can happen in 2 ways: In the first place it may be only the optical fixation reflexes in the nasalward direction that have developed further; in such cases the better eye will show a stronger jerking nystagmus in monocular vision than the worse eye; the optokinetic reaction will show greater differences with temporalward and nasalward contour movement for the better eye than for the worse eye. Examples of this are Cases 29, 30, 31, 32, 33, 34 and 35. In Cases 29, 30, 32 and 33 b — a was greater than c — d; in Cases 31, 34 and 35 b — a was less than c — d.

In the second place it is possible that for the better or more used eye both the temporalward and the nasalward optical fixation reflexes have achieved better development. But a relative excess of nasalward fixation tonus from the better eye may still persist, although it is also possible for this to disappear entirely (R.E. of Case 39). The result is that now the worse eye shows a stronger jerking nystagmus in monocular vision and that the optokinetic reaction to nasalward and temporalward contour movement is slighter and more abnormal for the bad eye than for the good eye. Examples of a development of this kind are Cases 36, 37, 38, 39, 40 and 41. The ideal result of such a development would be that despite the difference between b and d or between a and c, the relationship b — a = c — d had been achieved. In some of our cases it will be noted that transitions between this and the previous group exist; but however this may be, we have come to the conclusion that in all these cases the optical fixation reflexes from the better eye had advanced further in development.

In the examination of our patients with nystagmus we found a number in whom the occlusion of the worse eye led to only a very slight jerking nystagmus of the better eye, whereas occlusion of the better eye evoked a very lively jerking nystagmus of the worse eye. This puzzle was solved when we noted that all these patients (in so far as they were examined in this respect) showed a jerking nystagmus in the dark, with the fast phase in the direction of the worse eye. They thus had an asymmetric non-optical gaze tonus with a predominance to the temporal side of the better eye. In this way the excess of nasalward optical fixation tonus of the better eye was more or less compensated, whereas that of the worse eye was reinforced. All these cases have been placed in group V. The reactions to optokinetic stimulation in such cases are as follows: as far as the better eye is concerned a contour movement in the temporal direction will, with the aid of the non-optical gaze tonus, weaken, cancel or even overcome the existing jerking nystagmus (Cases 44, 45, 46 and 48); a contour movement in the nasal direction will for the better eye be opposed by the non-optical gaze tonus so that the existing jerking nystagmus will be little influenced or in some cases with a low general fixation tonus even weakened; for the worse eye a contour movement in the temporal direction will, opposed by the non-optical gaze tonus, be unable to do more than slightly weaken the existing jerking nystagmus, but will further leave it uninfluenced or in a few cases may even enhance it somewhat owing to exhaustion of the temporalward fixation tonus; a contour movement in the nasalward direction will, with the aid of the non-optical gaze tonus, practically always enhance the existing jerking nystagmus to a considerable degree for the worse eye.

We have thus become acquainted with 3 forms of asymmetric gaze tonus. These were observed in 20 patients, while in 15 cases there was reason to believe that the gaze tonus, both optical and non-optical, was practically symmetrical.

C. One-eyed patients with latent nystagmus.

Our 55 patients with nystagmus included 7 in whom one eye had been removed or was practically blind. In a few instances a slight pendular nystagmus was seen in daylight, while 3 patients (Cases 50, 53 and 55) showed a jerking nystagmus from time to time, with the fast phase in the direction of the sighted eye. When the sighted eye was covered, and also in the dark, all these patients showed a very definite jerking nystagmus with the fast phase in the direction of the absent or blind eye. There was thus undoubtedly an excess of non-optical gaze tonus in a direction temporal to the seeing eye. In daylight this excess was totally cancelled out by an asymmetric optical fixation tonus. We propose to devote a separate discussion to these important cases. At this point we shall only draw attention to their reactions to optokinetic stimuli. It is obvious that these cases are closely related to those of the last form of the asymmetric gaze tonus discussed above. With a contour movement temporalwards there appeared, aided by the non-optical gaze tonus, a marked jerking nystagmus in the nasal direction in all cases. With a contour movement in the nasalward direction, where the non-optical gaze tonus acted in opposition, we saw either a weak jerking nystagmus temporalwards or a weak jerking nystagmus nasalwards (inverse type); the latter indicates a low fixation tonus in the nasalward direction as well. In 2 cases (Cases 53 and 54) a very important observation was made. With the nasalward contour movement we saw first a weak jerking nystagmus temporalwards, after some time the eye ceased to move and finally an inverse type appeared, i.e. a nasalward jerking nystagmus. We regard this observation as giving important support to the view that the inverse type is due to exhaustion. We should further like to point out how these patients showed once more that a reinforcement of an existing optical fixation tonus is achieved with more difficulty than an inhibition thereof. The eye or eyes (in cases with an artificial eye this also moved) were kept in equilibrium by an excess of temporalward non-optical gaze tonus and an excess of nasalward optical gaze tonus. With a temporalward movement of the contours we saw a well-marked jerking nystagmus, which must be ascribed to an inhibition of the nasalward optical fixation tonus; with nasalward movement of the contours we saw only a very weak jerking nystagmus temporalwards or an inverse type; the only way in which this can be accounted for is by assuming that in some cases there was a very slight increase of the nasalward optical fixation tonus and in others a rapid exhaustion of it.

Summary.

The majority of patients with pendular nystagmus have a greatly reduced sensitivity to optokinetic stimulation, both with binocular and with monocular vision. With monocular vision an inverse type was observed in a few cases, especially with temporalward movement of the contours. A normal optokinetic jerking nystagmus, on the other hand, was more often seen with nasalward movement of the contours. All signs point in these cases to very weak optical fixation reflexes, while in some cases it appears that the temporalward optical gaze tonus in each eye separately was still worse than the nasalward.

A normal optokinetic jerking nystagmus in patients with pendular nystagmus or latent nystagmus, either with binocular or with monocular vision, may be regarded as exceptional. This again supports the hypothesis that these forms of nystagmus and the anomaly of the optokinetic nystagmus have a common cause, i.e. a disturbance in the development of the optical fixation reflexes. Only in cases with a pendular nystagmus in which the optokinetic reactions are normal is it not possible to ascribe this pendular nystagmus to a severe disturbance of the optical fixation reflexes; in such cases it is necessary to consider a deficiency in the light tonus.

In the majority of cases of latent nystagmus, optokinetic stimulation with binocular vision gave practically equal reactions for contour movement to the R and to the L. A symmetrical reaction of this kind is not as such proof of a symmetrical development of the optical fixation reflexes in the 2 eyes. An asymmetric reaction, however, does prove that the development of these reflexes is asymmetric The nature of the reaction to optokinetic stimulation with binocular vision may in itself give an indication as to how the optical fixation reflexes have developed in a given case. The optokinetic nystagmus of patients with latent nystagmus using one eye will practically always be different according to whether the contours are moved in the temporal or nasal direction. In the few cases where this reaction is equal for both directions it is necessary to consider the possibility of an asymmetric development of the optical or non-optical tonic innervation. A symmetrically equal reaction in both eyes indicates a symmetrical development of the optical fixation reflexes.

In cases of latent nystagmus it appears that a temporalward contour movement can more easily weaken the existing jerking nystagmus (48 eyes) than a nasalward contour movement can reinforce it (33 eyes).

As a general rule the optokinetic nystagmus has to give up the struggle against the jerking nystagmus of latent nystagmus. Where the jerking movement of latent nystagmus was vanquished by a temporalward contour movement (in 11 of the 70 eyes) we were obliged to ascribe this either to a very small difference between the nasalward and the temporalward fixation tonus or to an excess of non-optical gaze tonus in the temporalward direction.

An inverse type is seen only with a very low fixation tonus in the direction of the contour movement. An inverse type with contour movement in the nasalward direction can be promoted by an excess of non-optical gaze tonus in the temporalward direction.

An absence of effect of the optokinetic stimulation was very frequently seen in patients with latent nystagmus: 15 times with temporalward contour movement and 25 times with nasalward contour movement. This insensitivity to optokinetic stimuli also indicates a low level of optical fixation tonus. If in a patient with latent nystagmus one finds no reaction with nasalward contour movement but a reaction with temporalward contour movement, the chance is very great that this patient has an asymmetric gaze tonus.

The examination of optokinetic nystagmus is eminently suitable for the detection of asymmetric optical and non-optical gaze tonus. In this way we were able to distinguish 3 forms of asymmetric gaze tonus: (1) The development of the nasalward fixation tonus may be further advanced in the better eye than in the other eye. (2) Both the temporalward and the nasalward fixation tonus from the better eye may have developed further than those from the other eye. (3) An asymmetric non-optical gaze tonus may have developed, leading to compensation of the deficiency of temporalward optical gaze tonus from the better eye. Transitions between these 3 forms of asymmetry are also possible.

An important group is that of the one-eyed patients with latent nystagmus. In the dark they show a jerking nystagmus in the direction of the blind eye; in daylight they sometimes show a suggestion of jerking nystagmus in the direction of the good eye. With temporal contour movement these patients get a lively jerking nystagmus nasalwards and with nasal contour movement at the most a weak jerking nystagmus temporalwards, sometimes a cessation of movement of the eye and sometimes an inverse type. On 2 occasions we saw with nasalward contour movement a weak jerking nystagmus temporalwards which gradually changed over to a jerking nystagmus nasalwards (inverse type).

The manner in which the patients with pendular nystagmus and latent nystagmus reacted to optokinetic stimuli showed us with a fairly high degree of certainty that there must be a disturbance in the optical fixation reflexes. The manner in which the optokinetic nystagmus is manifested in such patients depends on (1) the severity of the disturbance in the development of the optical fixation reflexes and (2) the symmetry or asymmetry of the optical and of the non-optical reflexes.

V. Nystagmus in the dark.

Several of our 55 patients with nystagmus also showed a nystagmus in the dark. We had gained the impression that patients who had a pendular nystagmus in daylight also kept this in the dark, while patients with latent nystagmus did not have nystagmus in the dark if their eyes remained at rest in daylight. This impression was perhaps correct on the whole, but many exceptions were found. Unfortunately our original impression had the consequence that at first we did not regularly note whether a nystagmus could also be detected in the dark. This is the more regrettable as it later appeared that the existence and more especially the nature of a nystagmus in the dark could give very valuable information in connection with the understanding of the developmental disturbance responsible for the nystagmus.

For coarser movements it is possible to detect a nystagmus in the dark by palpation of the closed eyes. Finer movements cannot be detected in this way; the best way to observe them is to illuminate each eye dimly from the temporal side with an electric ophthalmoscope. We never saw any difference with illumination of the R.E. and the L.E., from which we conclude that illumination of the sclera had no influence on the nystagmus.

It is quite understandable that patients who have a pendular nystagmus in daylight should still keep it in the dark. Nystagmus is due to a deficient gaze tonus; this may be a deficiency of non-optical gaze tonus, a deficiency of optical gaze tonus or a deficiency of both these forms of gaze tonus. In the dark there is no reason whatever for the non-optical gaze tonus to increase or decrease, while the optical gaze tonus will, if anything, decrease. We saw only one patient (Case 1) belonging to group I (pendular nystagmus) who did not show nystagmus in the dark. But this patient did not really have nystagmus in daylight either, although she had formerly had it. She did, however, show a number of phenomena which were reminiscent of her earlier nystagmus, i.e. of a low gaze tonus. Her optical fixation reflexes were only slightly disturbed.

Many more exceptions to the above-mentioned impressions were presented by the patients with latent nystagmus. In the first place there were 4 patients (Cases 22, 27, 34 and 41) who showed no pendular nystagmus in daylight but did show it in the dark. In all these 4 cases we were obliged for various reasons to conclude that the optical fixation tonus was very low or the conjugated optomotor reflexes very poorly developed. We think it probable that in daylight the optical fixation tonus was still just capable of keeping the eyes at rest, whereas in the dark the absense of optical fixation reflexes gave free rein to the pendular nystagmus. A normal person does not have nystagmus in the dark, but the optical reflexes cannot act in the dark in the normal subject either. It is therefore the question whether our patients with pendular nystagmus in the dark did not also have a disturbance in the development of the non-optical gaze tonus.

It must in fact be assumed that such a developmental disturbance was present, but we believe it to be of secondary nature. In congenital blindness of both eyes, these eyes also do not remain at rest; they show the so-called nystagmus of the blind, which consists of irregular, often wandering and sometimes dissociated movements. Although we cannot call this pendular nystagmus, it does prove that the non-optical gaze tonus is then insufficient to keep the eyes at rest. In babies with papilla grisea who are born blind we also see these irregular eye movements at first; if the organ of vision begins to function so that the optomotor reflexes develop, there is always a stage during which these children show a true nystagmus, which may in the course of time disappear entirely, also in the dark. From this sequence of events we learn that the non-optical reflexes without the aid of optomotor stimuli can never become strong enough to keep the eyes at rest. Thus the optomotor stimuli to some degree enhance and regulate the non-optical gaze tonus also. This is not so very surprising in the light of our hypothesis that the conjugated optomotor reflexes are grafted onto the subcortical conjugated non-optical reflexes. By activating or inhibiting the subcortical reflexes, the cortical reflexes have taken these into their service.

An important indication that more attention ought to be paid to nystagmus in the dark was the fact that in a few patients with latent nystagmus we found a jerking nystagmus in the dark. We were able to detect this in 8 of the 35 patients (Cases 33, 37, 42, 43, 44, 45, 46 and 47). Where the visual acuity of the 2 eyes was unequal, the jerking nystagmus always had its fast phase in the direction of the worse eye; the only exceptions to this rule were cases 33 and 37.

A jerking nystagmus in the dark can only be due to an asymmetric non-optical gaze tonus. With latent nystagmus an asymmetry of this kind must have the consequence that the jerking nystagmus in monocular vision will be much stronger with one eye than with the other. We did in fact observe, in monocular vision with the better eye, either only a very weak temporalward jerking nystagmus or no nystagmus or a pendular nystagmus. Never, however, was the asymmetry of the non-optical gaze tonus so great that in monocular vision with either eye the fast phase of the jerking nystagmus was directed towards the heterolateral side. From this it may be concluded that an asymmetric non-optical gaze tonus can completely or partially abolish the jerking movement of latent nystagmus but can never prevail over it. This makes it highly probable that the asymmetry is not primary but has developed secondarily under the influence of corrective adjusting impulses from the preferred eye.

This is shown still more clearly by the one-eyed patients with latent nystagmus. All of them showed a pronounced jerking nystagmus in the dark, with the fast phase in the direction of the blind eye. In daylight there was either no nystagmus, a pendular nystagmus or a nystagmus with fine jerks, but now in the direction of the good eye. Thus in these patients also there was never a predominance of the non-optical gaze tonus in daylight, even though in the dark this gave such a marked jerking nystagmus in the direction of the blind eye.

Summing up we may thus make the following statements: (1) Where there is a pendular nystagmus in daylight, this is practically always present in the dark too. (2) Pendular nystagmus in the dark only and not in daylight is indicative of a low non-optical gaze tonus, which is probably secondary to a disturbance in the development of the optical reflexes. (3) Jerking nystagmus in the dark is due to an asymmetric non-optical gaze tonus, which has also almost certainly developed secondarily to compensate for the asymmetric optical fixation tonus of the preferred eye.

VI. Nystagmus and binocular perception.

In the examination of patients with strabismus it is often the custom to ascertain also whether there is a so-called normal or abnormal retinal correspondence. Normal correspondence is assumed to be present if the two foveal images combine to give a single perception and thus are localized at the same place; an abnormal correspondence is assumed to be present if the foveal image of the one eye is localized at the same place as a more peripherally situated retinal image of the other eye and can under suitable conditions combine with this to give one perception.

This correspondence can be examined by first causing one eye to fixate the middle of a vertical line of light, then causing the other eye to fixate the middle of a horizontal line of light and finally ascertaining whether the after-images of these 2 lines of light together form a cross. In this way the relative localization of the stimuli which impinge on the fovea and its immediate neighbourhood in the 2 eyes can be examined. One can also make use of the rods of Maddox, with a tangent scale, and ascertain whether the localization of the vertical line of light on the tangent scale corresponds to the angle of strabismus. In this way we ascertain to which part of the retina of one eye the fovea of the other eye corresponds.

Difficulties arise, however, if the image of one of the eyes, or sometimes of each in turn, is constantly suppressed; there is then no correspondence to be detected. It might indeed be said that in such cases no correspondence exists, but one has the uncomfortable feeling that the word 'correspondence' is not really suitable here. What can be said is that in these cases there is no functional junction between the perception of the right and that of the left eye; they cannot fuse to give a single entity or form. This functional junction does exist in cases where the 2 retinal images are perceived together and do fuse to give a single picture, irrespective of whether a so-called normal or abnormal correspondence is present.

Partly on the grounds of the relative positions of the cortical representations of the 2 retinae, it is highly probable that this functional junction also has an anatomical substratum, so that we may speak of a cortical binocular junction. On the grounds of the above considerations we think it better to drop the rather meaningless expression 'retinal correspondence' and to speak in future of cortical binocular junction (Keiner, 1951); this may be normal or abnormal but may also be entirely absent.

Must it now be accepted as a fact that in the absence of a cortical binocular junction the image of one of the eyes is constantly suppressed? Certainly not, for in that case a cortical binocular junction could never develop. Suppression also must, as it were, be learned, owing to the fact that throughout our lives we are continually suppressing images that do not fit into the pattern of our expectations (Ridley 1952). One comes across patients in whom a simultaneous perception can be evoked with a certain amount of difficulty, but who are not able to fuse these simultaneous perceptions into a single picture and who have the greatest uncertainty in estimating the subjective distance between 2 monocular perceptions. In such cases we feel justified in concluding that there is no cortical binocular junction.

Concomitantly with the cortical junction there develops also, in our opinion, the convergence innervation from the double adduction innervation. Therefore also the cortical binocular junction is not a point to point junction but the cortical representation of each retinal element is connected with a larger or smaller number of cortical representations of the other retina (Panum's perception circle or fusional area). The wider this area the more flexible is the binocular junction, the greater the fusion amplitude and the better the development of depth perception. It is obvious that according to this view the convergence innervation and everything associated with it has originated from a further development of the monocular optomotor reflexes.

If we now return to our nystagmus patients and enquire how matters stood with their cortical binocular junction, depth perception, fusion amplitude and suppression, we find the following facts: Of the 48 patients with pendular nystagmus and latent nystagmus, 2 were too young and one too mentally defective for us to be able to obtain reliable information as to binocular perception. Of the other 45 patients, we were unable to find evidence of cortical binocular junction in 23, while we found a more or less developed normal binocular junction in 20 and an abnormal binocular junction in 2. Among the 23 patients without cortical binocular junction there were 3 in whom it was possible to elicit a simultaneous perception, but without the possibility of fusion of the 2 retinal images, while suppression of one of the images also occurred very readily. These 3 were Cases 25, 32 and 46. Among those who lacked cortical binocular junction were 6 patients with pendular nystagmus and 17 with latent nystagmus.

Among the 20 patients with a more or less developed normal cortical binocular junction, 5 with pendular nystagmus and 15 with latent nystagmus, there were 13 with a certain degree of depth perception and 7 with binocular fusion but no depth perception. We speak of a certain degree of depth perception because gradual transitions were seen. The examination was carried out with stereoscope plates. There were some patients with very good stereoscopic perception, but there were some who, although they immediately recognized the correct depth relationships with easy stereoscope plates, recognized difficult ones only after sometime or not at all. A very remarkable phenomenon that we encountered several times was the following: If one displays to the 2 eyes 2 figures of equal size, one of which is seen at a somewhat greater convergence angle than

Two of the most frequently used stereoscope plates (drawn from existing plates)

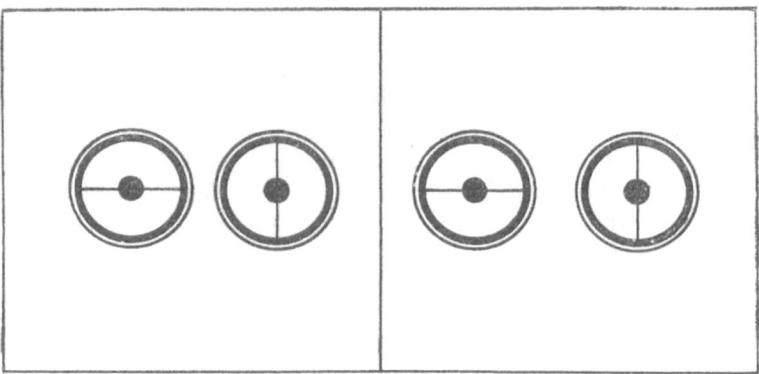

A. An easy stereoscope plate, in connection with which one can ask the following questions:
    (1) which circle lies further in front, the one with the horizontal or the one with the vertical spokes?
    (2) which circle is apparently the smaller?

The correct answer is that the circle with the horizontal spokes is apparently closer to the eye and apparently smaller.

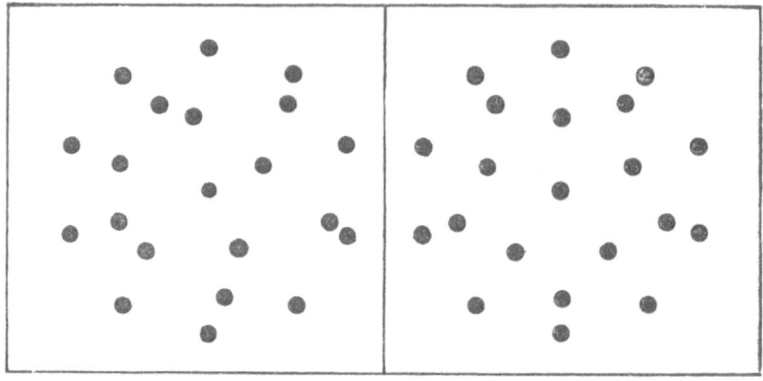

B. A difficult stereoscope plate, in connection with which one can ask the following questions:
    (1) How are all the dots arranged?
    (2) Where is the middle dot placed?

The correct answer is that the dots are arranged in 3 planes, 5 in the front plane, 11 in the middle plane and 5 in the back plane; the middle dot lies in the middle plane.

the other, the first figure will appear to be not only closer but also smaller. A few patients (Cases 15, 22 and 27) were able to recognize the apparent difference in depth but not the apparent difference in size. This phenomenon — to which we did not pay attention in all cases — confronts us with the question of whether the apparent difference in size with retinal images of equal sizes can really be dependent only on the apparent distance (reference surfaces of Miles, 1951).

We found a very good correlation between the degree of depth perception and the fusion amplitude. We speak of a very small fusion amplitude when this does not amount to more than a few degrees; a small fusion amplitude when it is not more than $15°$ (in most cases it was less thans this) and a good fusion amplitude when it is more than $15°$. Table V gives a good impression of the correlation between the degree of binocular perception and the fusion amplitude.

TABLE V

	no fusion	very small fusion amplitude	small fusion amplitude	good fusion amplitude
simult. without fusion	3	0	0	0
simult. with fusion	0	5	1	1
fusion with depth perception	0	2	8	3

Correlation between degree of binocular perception and fusion amplitude

This correlation is entirely in accordance with our theoretical considerations as described in the foregoing. On these grounds it was also to be expected that there would be a correlation between the fusion amplitude and the more or less prompt occurrence of suppression of one of the retinal images. There were some patients in whom it was almost or quite impossible to evoke double images (very rapid suppression) and others in whom double images could be made visible only with difficulty or by the use of tricks (quick suppression) and others again in whom double images were easily evoked and also persisted (no suppression). Table VI gives a survey of the correlation between the fusion amplitude and the degree of suppression.

With a very small fusion amplitude the suppression was now and then so prompt that it was difficult to make out whether binocular fusion still existed or suppression had already taken place.

TABLE VI

	no fusion	very small fusion amplitude	small fusion amplitude	good fusion amplitude
very quick suppression	3	7	3	0
quick suppression	0	0	4	0
no suppression	0	0	2	4

Correlation between degree of suppression and fusion amplitude.

And now a few words about our 2 patients with an abnormal cortical binocular junction (Cases 18 and 30). Case 18 was particularly remarkable. This patient had only a very slight convergent squint of the R.E., which could be easily detected by means of the adjusting movement of the R.E. when the L.E. was occluded. But with Maddox' rods the result always corresponded to a slight exophoria. Despite this abnormal binocular junction she did have some degree of binocular depth perception, although the difference in depth was not always perceived immediately and the fusion amplitude was small, while the stereoscope plates had to be presented in the correct way, otherwise one of the retinal images was very quickly suppressed. Case 30 had strabismus convergens of the left eye with slight alternating hyperphoria. She showed clear evidence of the abnormal junction and had — as is the case with the majority of patients with abnormal binocular junction — simultaneous perception with fusion of the 2 retinal images but no depth perception. She had a very small fusion amplitude and prompt suppression of one image.

It may well be asked why an abnormal junction was so seldom found in our nystagmus patients with strabismus. The reason is probably the same as that for the fact that in many of the nystagmus patients without strabismus the binocular perception was defective with a too small fusion amplitude. If with a normal position of the eyes the monocular optomotor reflexes do not evoke a normal cortical binocular binding, the formation of an abnormal cortical binocular junction in strabismus patients will be a great deal more difficult still, because the cortical representations of the retinal elements concerned occupy less favourable anatomical positions with respect to each other. The question still remains, of course, whether the nystagmus is primary and has impeded the development of the monocular reflexes or whether the poor development of the monocular optomotor reflexes is primary and has favoured the occurrence of nystagmus. Nystagmus is not in every case the consequence of a disturbance in the monocular optomotor reflexes, certainly not in the case of latent nystagmus.

Summing up the results of our investigation of binocular perception we come to the following conclusions:

(1) Both in pendular nystagmus and in latent nystagmus there is very often no detectable cortical binocular junction. This was the case in about half our patients.

(2) The normal cortical binocular junction, although in many cases detectable, is, in patients with nystagmus, often far from ideal (this last was seen only in Case 11, 28 and 35). A large proportion of patients showed defective depth perception, too small fusion amplitude and too rapid suppression of one of the retinal images.

(3) An abnormal cortical binocular junction was a great rarity among our nystagmus patients (only in Cases 18 and 30). Even without strabismus the development of the normal binocular junction was mostly defective; with strabismus the development of the abnormal binocular junction must be still more difficult, on account of the less favourable anatomical relationships.

(4) Since we ascribe a good development of visual acuity and of binocular perception to a good development of the monocular reflexes, we are obliged to assume that in the great majority of our nystagmus cases the development of the monocular optomotor reflexes was disturbed to a greater or less degree.

VII. **Optical localization in patients with nystagmus.**

In 1928 Roelofs investigated optical localization in 3 patients with very marked latent nystagmus. Using a fixed fixation point in monocular tests, all 3 patients localized an object in the field of vision more to the right with the R.E. and more to the left with the L.E. With the eyes moving in monocular tests, an object in the temporal part of the field of gaze was judged to be further away than an object at the same distance from the centre in the nasal part of the field. All this appeared to indicate a predominance of the tonic innervation in the nasal direction in monocular vision.

Esters (1930), who repeated this investgation on a number of patients with latent nystagmus, was unable to confirm the above results. In 14 of our patients we subsequently determined the position of the optical median plane, both in binocular and in monocular vision. The results showed that Esters was right. In some cases the optical median plane was deviated nasalwards in monocular vision, while in other cases it was deviated temporalwards.

We take the view that a point of light will be localized in the median plane of the head if no innervation impulse to turning to the right or to the left is needed for the fixation of that point. If we find, thus, that in latent nystagmus the optical median plane is displaced to temporal in monocular vision, we must assume that the impulses which give rise to the slow phase nasalward, which are almost certainly of optical nature, are slightly weaker than the corrective impulses which give rise to the fast phase, without the co-operation of optical adjusting impulses.

However, we have not analysed our results further, as they are also influenced by various factors involved in the optical localization of patients with strabismus, giving rise to all kinds of new problems which would lead us too far away from our present subject. We can only state that the influence of latent nystagmus on optical localization is also far from simple on account of accessory factors.

VIII. **Latent nystagmus in one-eyed persons.**

Our patients with latent nystagmus included 7 who only had the sight of one eye. On looking straight ahead they showed practically no nystagmus, but when the sighted eye was covered

— i.e. in the dark — they showed a marked jerking nystagmus with the fast phase in the direction of the blind eye. In cases where occlusion of one eye gives a jerking nystagmus in the direction of the non-occluded eye, one immediately thinks of latent nystagmus. In these cases, however, occlusion of the good eye amounted to the same as occlusion of both eyes, so that one rightly hesitates to place these patients in the group of latent nystagmus cases. On the grounds of sufficient arguments, however, we consider that such a classification is justifiable. Not only the remarkable phenomena observed but also and more especially the important conclusions resulting therefrom, have induced us to devote a more detailed discussion to the monoculi with latent nystagmus.

We shall begin by mentioning a few cases from the literature and shall then proceed to describe our own cases.

First of all 2 cases of C. and H. Fromaget (1916):

The first patient had strabismus divergens. His right eye was totally blind as a result of an accident in childhood. The left eye was slightly myopic; the visual acuity was $^1/_2$; with correction 1. This eye did not show even the slightest quivering. But as soon as the good L.E. was occluded by a shield there appeared a rapid horizontal nystagmus with about 160 strokes per minute. Unfortunately it is not stated whether this was a jerking or a pendular nystagmus. We can say with certainty that this nystagmus was evoked by non-optical stimuli and suppressed by optical stimuli.

The second patient also had a right eye which was totally blind as a result of a perforating trauma, with strabismus divergens. The L.E. was emmetropic with a visual acuity of 1. Here again occlusion of the good L.E. evoked a nystagmus with large excursions and a frequency of about 60 per min. Whether it was a pendular or a jerking nystagmus is not stated in this case either. The authors assume, in connection with these cases and with the ordinary forms of latent nystagmus, that all the retinal stimuli which are transmitted to the tonic co-ordination centre are necessary for a good ocular equilibrium. In the cases described it is believed that owing to the loss of one eye the development of the co-ordination centre was retarded, hence the strabismus and the tendency to nystagmus. A similar effect is ascribed to all causes tending to reduce the quality and quantity of retinal stimuli (maculae corneae, cataract, retinal dialysis, optic nerve atrophy).

The above authors, however, overlooked the fact that in latent nystagmus cases there is often no nystagmus in the dark or on occlusion of both eyes, so that reduction of the light stimuli cannot be the sole cause. This is the more surprising in view of the fact that in the same publication they describe the case of a 24-year-old soldier who had developed after the loss of his right eye a nystagmus which disappeared when the good eye was covered.

Although the explanation offered by these authors cannot be valid for latent nystagmus in general, for the 2 cases described it did seem very likely. But if we knew that the nystagmus on occlusion of the good eye had been a jerking nystagmus, we should give preference to another explanation.

The third case from the literature that we should like to quote was described by Ohm (1928). The patient was a man aged 25 yr. with strabismus divergens of the R.E. This eye showed numerous anomalies and was practically blind. The left eye was normal and had a visual acuity of 1. When the patient looked straight ahead the good L.E. showed a fine rotatory nystagmus with very fine horizontal jerks to the left. When the good L.E. was occluded there appeared a lively jerking nystagmus to the R. while the nearly blind R.E. turned about 3 mm. downwards and rolled slightly to the left (endorotation) and the left eye showed a corresponding exorotation but no vertical movement. When both eyes were closed there was also a jerking nystagmus to the right; when the eyes were turned to the left this decreased and sometimes even changed over to a jerking nystagmus to the left. Illumination of the closed eyelids had little effect. When the left eye was illuminated from the R. or from the L. with a small pencil of rays in twilight, the jerking nystagmus to the R. persisted, but it stopped as soon as the rays impinged on the retina near the fovea.

Ohm regards this case as proof that visible contours are not necessary for the evocation of latent nystagmus. For this and similar cases there is little to be said against this view, but it must be pointed out that visible contours were necessary to inhibit the nystagmus; illumination of the closed eyelids did not suffice. It might be thought that this influence of contours had developed secondarily as a compensatory reflex, but if so it had overshot the mark, because when the eye was open there was a jerking nystagmus to the left, although this was only slight. On these grounds one is more inclined to surmise that the non-optical impulses have a compensatory significance which tends to correct a bad optomotor function.

Ohm sought the cause of these anomalies in an affection of a relay centre superior to the ocular muscle nuclei and situated in the ventro-caudal part of the nucleus of Deiters. He based this suggestion in the first place on the fact that similar forms of jerking nystagmus have been observed in affections of the vestibular nucleus and in the second place on the occurrence of a difference in height of the 2 eyes accompanied by rotation of both eyes. But there is a discrepancy here. He observed in his patient that a jerking nystagmus to the right was accompanied by a turning downwards of the right eye. An affection of the left vestibular organ also gives a jerking nystagmus to the right, but then according to Pötzl & Sittig (1925) we should expect a downward movement of the left eye or an upward movement of the right eye. A little further on we shall show how Ohm's accurate investigations and his detailed descriptions of the phenomena observed by him in his patient, which were so entirely in agreement with what we observed in our own cases, led us onto a different track.

The fourth case is taken from Crone's thesis (1952). The patient was a girl aged 17 years with on the right a micropthalmus and coloboma iridis, choroideae et nervi optici, in addition to cicatricial choroiditis. The visual acuity was limited to light perception ($^1/\infty$). The left eye was 3 D. hypermetropic with a visual acuity of 1 and there were no further anomalies. With both eyes open there was a fine jerking nystagmus to the left, sometimes with a slight rotatory component. The right eye was too high and executed slow, dissociated vertical movements. The hyperphoria of the right eye increased in abduction and decreased in adduction. When the right eye was occluded it turned up still further. When the left eye was occluded a slight jerking nystagmus to the right appeared and the right eye moved downwards to a marked degree.

The similarity to Ohm's case is very striking, except that the jerking nystagmus to the right was somewhat less lively in Crone's case.

To these 4 one-eyed patients with latent nystagmus we have been able to add 7 more (Cases 49, 50, 51, 52, 53, 54 and 55). Case 53 is also mentioned in Crone's thesis, but was examined by one of us as well.

Among the 7 monoculi were 2 with an artificial eye. In Case

49 the left eye had been enucleated at the age of 6 months on account of pseudoglioma; in Case 50 the left eye had been removed at the age of 3 years on account of microphthalmus. Microphthalmus was also the cause of the unilateral blindness of Cases 51 and 52; these patients had only light perception in the bad eye. In cases 53 and 54 an old dialysis retinae was the cause of the unilateral blindness, while in Case 55 extensive maculae corneae and anterior synechiae had reduced the visual acuity of the R.E. to $^{0.5}/_{300}$.

The patients who had lost one eye at a very early age and also those with unilateral microphthalmus show that latent nystagmus cannot be due to an abnormal cortical binocular junction, as in them all chance of development of any cortical binocular junction was a priori excluded.

If we also take the cases from the literature into consideration, we get the impression that among one-eyed patients with latent nystagmus there are more cases of strabismus divergens than of strabismus convergens. This, however, is not so very surprising, as no optomotor impulses whatever emanate from the blind eye and this will therefore tend to assume its anatomical resting position.

A fact which may be important is that a few of the monoculi with latent nystagmus showed a suggestion of alternating hyperphoria. This was the case in the patients of Ohm (1928) and Crone (1952) and in 2 of our own patients (Cases 53 and 55). Case 50 had a rotatory pendular nystagmus on looking straight ahead, but owing to the fact that she wore an artificial eye it was not possible to ascertain whether alternating hyperphoria was present.

According to Crone, and in our opinion also, alternating hyperphoria is due to a primary disturbance in the development of the monocular optomotor reflexes. Latent nystagmus also belongs to the syndrome of alternating hyperphoria. Crone succeeded in finding a convincing explanation of the elevated fixation tonus in the nasalward direction in monocular vision in patients with alternating hyperphoria. This increased optical fixation tonus in the nasalward direction causes the jerking nystagmus. As we shall see, we also found evidence of an increased optical fixation tonus in the nasalward direction in the latent nystagmus of one-eyed patients, and it seems very likely that this too must be regarded as a primary disturbance.

The insufficiency of the optical fixation reflexes in the temporalward direction was shown up very clearly when we made our one-eyed patients gaze temporalwards and nasalwards. Table VII gives a survey of this.

TABLE VII

Case No.	Gaze straight ahead	Gaze temporalwards	Gaze nasalwards
49	rest or irregular fine movements	jerking nyst. temp.	pendular nyst. with a few jerks nas.
50	rest or rotat. pendular nyst. with jerks temp.	rotat. jerking nyst. temp.	diagonal pendular nyst. with rotat. jerks temp.
51	at rest	jerking nyst. temp.	at rest
52	fine tremor	fine tremor	fine tremor
53	weak jerking nyst. temp.	stronger jerking nyst. temp.	very weak jerking nyst. temp.
54	not quite at rest	jerking nyst. temp.	jerking nyst. nas.
55	rest or jerking nyst. temp.	stronger jerking nyst. temp.	at rest

Nystagmus in sideways gaze of one-eyed patients with latent nystagmus.

With a single exception (Case 52) there appeared with temporalward gaze a marked jerking nystagmus temporalwards or a considerable accentuation of a weak pre-existent jerking nystagmus in that direction. With nasalward gaze, on the other hand, either the eye remained at rest or there was some degree of pendular nystagmus or very weak temporalward jerks were still visible or jerks in the nasalward direction appeared. The last-mentioned was seen in 2 cases (Cases 49 and 54) and showed that there was here also an insufficiency of the fixation reflexes in the nasalward direction.

If the good eye was occluded or the patient was placed in a dark room, a very marked jerking nystagmus with the fast phase in the direction of the blind eye was seen in all 7 cases. In some cases the placing of a strongly positive lens (spher. + 20) before the good eye, so that the surrounding contours became less distinctly visible, even sufficed to evoke this nystagmus. With the eyes closed, the jerking nystagmus could not be abolished by strong illumination through the closed lids of the good eye. The jerking nystagmus was a conjugated movement; where an artificial eye was worn, this was often seen to move with the sighted eye.

It follows from the foregoing that this nystagmus of one-eyed patients is due to non-optical reflexes and that in daylight it is suppressed by optical reflexes. These optical reflexes are not evoked by illumination of the eye but occur in response to the presence of visible contours. Moreover, since the predominance of the optical reflexes acts only in the nasalward direction, it is obvious that there is an asymmetry of the optical fixation reflexes. In the absence of any cortical binocular junction, the conjugated nature of the nystagmic movements in itself suffices to show that the disturbance must be sought in the reflex pathway for conjugated movement.

In our one-eyed patients with latent nystagmus we thus find on the one hand a predominance of non-optical fixation reflexes in the temporalward direction and on the other hand a predominance of optical fixation reflexes in the nasalward direction. This brings us to the question of which of these affections is primary and which secondary.

We consider that there is every reason to believe that the disturbance in the optical fixation reflexes is primary. In the first place the jerking nystagmus in temporalward gaze indicates that the fixation tonus in that direction is deficient. Thus instead of an excess of fixation tonus in the nasalward direction we should more correctly speak of a deficit of fixation tonus in the temporalward direction. This is the more so in view of the fact, as we shall see presently in connection with the reactions to optokinetic stimulation, that the fixation reflexes and the fixation tonus in the nasalward direction also left something to be desired. In the second place it must be pointed out that even when the good eye was open it did not always remain completely at rest, but when it did show nystagmus this had its fast phase not in the direction of the blind eye but in that of the good eye. This was seen in 3 of our cases and also in Ohm's case. This is quite understandable if the asymmetric optical tonic innervation is primary and the non-optical innervation is regarded as a compensation. It would be less understandable if the asymmetric non-optical tonic innervation were primary and the optical compensatory innervation had overshot the mark.

In the third place we consider it an important argument that some of the one-eyed patients showed signs of alternating hyperphoria, a phenomenon which in itself is evidence of a primary disturbance of the optomotor reflexes.

A feature which we considered particularly interesting was the behaviour of the eyes on optokinetic stimulation. We propose to discuss this in rather more detail.

If latent nystagmus is due to the fact that small contour movements over the retina in a temporalward direction do give an increase of fixation tonus but displacements in a nasalward direction do not, this implies that the fixation reflex in the nasal half of the retina has an inhibitory and not an activating influence on the fixation tonus and that in the temporal half of the retina it has an activating and not an inhibitory influence. Under these conditions, with monocular vision, a temporalward movement of contours to evoke optokinetic nystagmus (which is based on the fixation reflexes) would be able to counteract the latent nystagmus; an inverse type would then have to be ascribed to adjusting movements. A nasalward movement of contours would be able to reinforce the latent nystagmus to some degree, but since stimulation and inhibition are already at work, not much can be expected from such a reinforcement. Where the optical fixation reflexes are insufficiently developed, a reinforcement of this kind can also lead to exhaustion and abolish the optokinetic nystagmus. Under these conditions also, an inverse type can be evoked by adjusting impulses.

If now, as we believe to be the case in our patients, an asymmetric tonic innervation from non-optical stimuli has developed in a reflex manner by way of compensation, the state of affairs becomes somewhat different. A temporalward movement of the contours can cancel out the optical influences but not the non-optical influences; as a result of this a temporally directed predominance of the non optical impulses persists and we get — as with occlusion of the eye — a jerking nystagmus to the side of the blind eye. No abolition of the latent nystagmus is then to be expected and still less an inverse type. A nasalward contour movement can produce hardly any further increase in the already existing predominance of the optical tonic innervation in the direction of the blind eye, but with poor development there is a great chance of exhaustion, in which case an inverse type will inevitably appear.

It is practically certain that we may take it as a general rule that whenever the eyes are held in equilibrium by asymmetric optical and asymmetric non-optical reflexes a contour movement in the direction in which the optical tonus predominates will give only a weak reaction and a contour movement in the direction in which the non-optical tonus predominates will give a well-marked jerking nystagmus. From this we might deduce that on displacement of images over the retina the inhibition of fixation tonus has in general a greater influence than the activation thereof. Table VIII shows to what degree the examination of optokinetic nystagmus gave results in accordance with our expectations.

TABLE VIII

Case No.	Temporalward contour movement	Nasalward contour movement
49	eyes at rest for a moment; then coarse jerks nasalwards	eyes at rest for a moment; then fine jerks nasalwards
50	jerking nyst. nasalward	pendular nyst.; sometimes jerks nasalwards and upwards
51	strong jerking nystagmus nasalwards	weak jerking nyst. temporalwards
52	very strong jerking nystagm. nasalwards	weak jerking nyst. temporalwards
53	increasing jerking nyst. nasalwards	first jerking nyst. temporalwards; then rest; then jerks nasalwards
54	jerking nyst. nasalwards (small amplitude)	first pendular nyst.; then jerks nasalwards
55	jerking nyst. nasalwards	very weak jerking nyst. temporalwards

Optokinetic nystagmus of one-eyed patients with latent nystagmus.

It is clear that these results are fully in accordance with our expectations. With a contour movement temporalwards the excess of optical fixation tonus nasalwards is inhibited; the excess of non-optical gaze tonus temporalwards persists unchanged and the result is a jerking nystagmus nasalwards. With contour movement nasalwards the excess of non-optical gaze tonus in the temporalward direction also remains unaffected, while the excess of optical fixation tonus nasalwards can hardly be increased any further and thus soon becomes exhausted, the result being a very weak temporalward jerking nystagmus, a pendular nystagmus or an inverse type. In Cases 53 and 54 the onset of exhaustion could be clearly followed in the reaction to optokinetic stimulation.

It is obvious that a compensation by non-optical reflexes is to be expected particularly where a compensation by optical reflexes from the other eye is not possible, as is the case in one-eyed individuals. But one is also inclined to wonder whether in some cases of latent nystagmus where one eye is exclusively used, either because it is the preferred eye or because the sight of the other one is so bad, the conditions are not also such as to promote an adaptation of the non-optical reflexes. And this is indeed the case; these are the patients of our group V and they show many phenomena in common with those of group VI. Cases of this kind have also been described by Verhage (1941; 1942).

One of these phenomena is that in monocular vision with the better eye the nystagmus is only very weak and on occlusion of this eye a marked jerking nystagmus appears with the fast phase in the direction of the worse eye. If there is also a jerking nystagmus in the dark, the asymmetry of the non-optical gaze tonus is proved. If there is no jerking nystagmus in the dark, one must be cautious with the diagnosis. It is quite possible that for the better eye the optical fixation reflexes have developed more equally than for the worse eye; in monocular vision with the better eye there will then also be a less marked nystagmus. There is then an asymmetry of the optical fixation reflexes of the 2 eyes, which is sometimes recognisable from a crooked position of the head. Once again it is the reaction to optokinetic stimulation that can confirm this surmise. In this connection we draw attention to the case-histories of Cases 38, 39 and 40.

An excellent illustration of our hypothesis that the non-optical gaze tonus can sometimes play a compensatory role is given, in our opinion, by a case described by Ohm (1928). The patient, a man aged 32 years, had a strabismus divergens of the right eye. When the right eye was occluded he showed a very slight jerking nystagmus to the left; when the left eye was occluded a lively jerking nystagmus to the right appeared. Both eyes had a posterior polar cataract. The visual acuity of the R.E. was $^4/_{60}$ (with correction $^4/_{35}$) and that of the L.E. was $^4/_{35}$ ($^4/_{20}$ with correction). A cataract operation was performed on the left eye, as a result of which its visual acuity rose to $^4/_8$.

In order to ascertain whether the jerking nystagmus to the right would now gradually disappear, the left eye was kept covered by a bandage for 12 weeks. The jerking nystagmus to the right definitely decreased but did not entirely disappear. It was also found that with a diaphragm with a large aperture placed before the R.E. the jerks to the right were stronger than with a diaphragm having a small aperture before that eye; thus the increase of contours visible to the R.E. enhanced the jerking nystagmus to the right. This shows that the prolonged occlusion of the L.E. was a two-edged sword; on the one hand the compensatory predominance of the leftward non-optical impulses was weakened, but on the other hand the predominance of the leftward optical fixation tonus was accentuated. When the left eye was finally uncovered and began to fixate there was a jerking nystagmus to the left (owing to the loss of the compensatory non-optical gaze tonus), which decreased greatly in the course of time. To prevent misunderstanding we must point out here that the explanation given by us is not that of Ohm himself.

We are thus of the opinion that the asymmetry of the tonic gaze innervation by non-optical stimuli must be regarded in many cases as a compensation or adaptation. This is supported by the fact that the asymmetry is the most pronounced in patients who are blind in one eye. If the worse eye is not blind, an asymmetry in the optical fixation reflexes for that eye can develop, just as for the better eye; in such a case the non-optical fixation reflexes do not need to provide compensation, as the optomotor reflexes from the right and left eyes then balance each other.

And how must we now imagine the asymmetry in the non-optical reflexes to come about? It is probable that the fast phase is due partially or chiefly to non-optical impulses. If now the optical fixation tonus in the nasalward direction predominates and if the patient always looks with the same eye — as is the case par excellence in persons with one eye — the fast phase will always be in the same direction and the non-optical reflexes will always send out their impulses in the same direction. This continual stream of impulses will gradually raise the non-optical tonic innervation in the direction of this fast phase, so that the higher optical tonus in the opposite direction is more or less compensated (elevation of tonus by summation of stimuli).

Nevertheless, the tendency to compensation by non-optical impulses is subject to wide individual variation; this may perhaps be connected with the degree to which the fast phase of latent nystagmus is produced by non-optical or by optical impulses. We shall now quote a few examples of such individual differences from the literature:

First we would draw attention to a case of C. & H. Fromaget (1916). A soldier aged 24 yr. had lost his right eye in 1914. With the left eye open he showed a nystagmus both of the left eye and of the stump remaining on the right. When the left eye was closed the nystagmus disappeared in 5—10 seconds. Apparently in this case the adaptation had wholly or partially failed to take place.

Secondly we wish to mention Ohm's patient (1928) to whom we have already referred. In this case occlusion of the left eye for 12 weeks led to a partial improvement in the jerking nystagmus in monocular vision with the right eye. Here, thus, there was some adaptation. It must be taken into consideration, however, that 12 weeks is a rather short time, particularly in such a case in which a predominance to the left of the non-optical impulses had already developed.

More important is perhaps another case described by Ohm in 1942. This patient was a man aged 27 years who had lost his right eye at the age of 13 years. On looking straight ahead the eyes were sometimes at rest, but there was often a jerking nystagmus to the left. When he looked to the left this jerking nystagmus increased and when he looked to the right it disappeared and was finally replaced by a jerking nystagmus to the right. When the left eye was open there was thus a slight predominance of the gaze tonus to the right. When the left eye was occluded a jerking nystagmus to the right appeared, this being also shown by the artificial eye on the right.

It appears likely that with the eye open this patient had an excess of gaze tonus to the right and with the eye closed an excess of gaze tonus to the left. The excess with the eye opened must be ascribed to optical stimuli and that with the eye closed to non-optical stimuli. A complete compensation by the non-optical stimuli had admittedly not be achieved, but compensation was unmistakably present. Ohm will have difficulty in accepting our explanation. He writes as follows: 'The influence of darkness is of the greatest importance for the theory of latent nystagmus. Here neither the fixation reflexes (Kestenbaum) nor the musculo-sensory reflexes (Roelofs) are involved.' We, however, are precisely of the opinion that the excess in the optical impulses is due to the optical fixation reflexes (Kestenbaum) and the excess in the non-optical impulses partly to the musculo-sensory reflexes (Roelofs).

It appears that the adaptation by means of non-optical impulses may also vary; this at any rate could be deduced from the phenomena in the case of a 14-year old boy whose case was described by Sorsby (1931). At the age of 10 yr. he had shown a lively nystagmus when the right eye was occluded; when he was 13 it was found, however, that nystagmus of the right eye appeared when the left eye was occluded and that the left eye showed a tendency to nystagmus when the right eye was covered. When he was 14 the right eye remained still when the left eye was occluded and the left eye showed nystagmus when the right eye was occluded. This boy had a convergent strabismus of the left eye. There was probably an excess of non-optical gaze tonus to the right, which had been temporarily lost at the age of 13 years.

If we now consider our own cases of group V in which we believe a compensatory tonic innervation by non-optical impulses to be present, we find that the adaptation is not equally great in all cases. Arranged according to the degree of adaptation, beginning with the most nearly complete adaptation, they fall into the following order: Case 48, 47, 46, 45, 44, 43, 42.

We may thus sum up our results as follows:
(1) There are some one-eyed patients in whom the occlusion of the good eye or even only the absence of visible contours evokes a jerking nystagmus with the fast phase in the direction of the blind eye.
(2) This nystagmus must be maintained by an asymmetric non-optical tonic innervation.
(3) The presence of visible contours causes this jerking nystagmus to disappear or to be transformed into a slight jerking nystagmus in the opposite direction.
(4) The assymmetries of the tonic innervation by non-optical impulses and that by optical impulses more or less balance each other in these one-eyed patients.
(5) The predominance of the optical asymmetry and the frequent occurrence of traces of alternating hyperphoria render it probable that the optical asymmetry is primary and the non-optical asymmetry a secondary compensation. In any case the marked asymmetry of the tonic innervation by non-optical impulses has something to do with the fact that the patient only has the sight of one eye.

(6) In patients with 2 functioning eyes the asymmetric optical tonic innervations from each eye will more or less balance each other. If they fail to do so, it can be shown with a high degree of probability in some cases that here also a compensatory tonic innervation by non-optical impulses has been established.

(7) The suggestion of alternating hyperphoria in some of our one-eyed patients shows that in addition to the disturbance in the reflexes for conjugated movements, as deduced from the latent nystagmus, there must also be a disturbance in the monocular optomotor reflexes.

IX. Gaze tonus and nystagmus.

The foregoing has, we believe, made it sufficiently clear that nystagmus is the consequence of an abnormal gaze tonus and that in very many cases this abnormal gaze tonus is a consequence of a disturbance in the development of the optomotor reflexes. As we have already pointed out, it can be deduced from the tonus of the ocular muscles that the co-ordination centres for eye movements are in a continuous state of excitation. This continual excitation state is maintained in a reflex manner by numerous stimuli of very different kinds which reach the co-ordination centres, to a large degree via subcortical pathways but to some degree also via cortical routes. Impulses to voluntary eye movements such as adjustment movements and shifting of the direction of gaze are excluded here; although one may speak of adjustment reflexes and direction reflexes they are of a different character; they lack the characteristics of continuity and tonicity and lie at a higher level in the central nervous system. The reflex continuous excitation state can be called innervation tonus, or where the gaze movements are concerned the term gaze tonus (Roelofs 1937) can be used.

The gaze tonus is maintained by stimuli of non-optical and of optical nature. The non-optical stimuli include vestibular, musculo-sensory and sympathetic stimuli. Among the stimuli of optical nature one can distinguish those which follow a reflex path that has been grafted onto or calibrated by the proprioceptive reflex paths and which originally gave rise to monocular reactions, and those which follow a reflex path that has been grafted onto and calibrated by the vestibular reflexes and has therefore given rise from the first to conjugated eye movements.

As a result of this calibration and by further autonomous development, each part of the retina — and in the macular region actually each individual retinal element — will send out its own graduated optomotor impulse, so that in vision with the whole retina a tension pattern is formed, the resultant of which contributes to the gaze tonus.

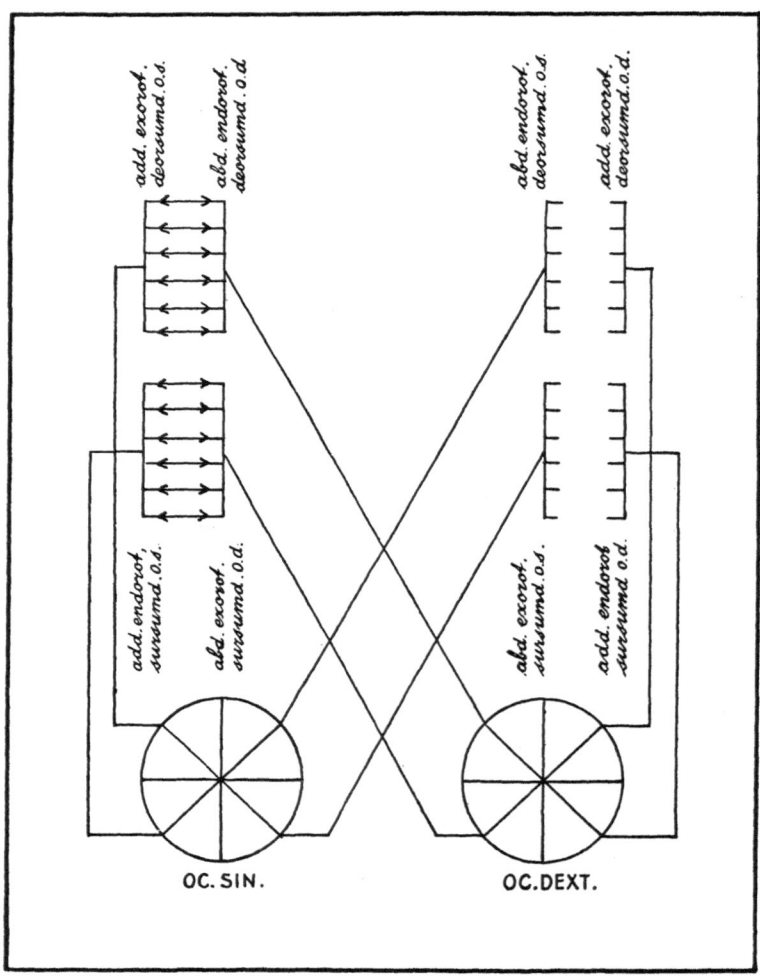

A graphical and schematic representation of the monocular optomotor reflex paths from the individual retinal quadrants to the occipital cortex. In the left half of the brain the cortical binocular junction of these monocular reflexes is shown; in the right half this junction is absent.

In virtue of their origin the monocular optomotor reflexes will strive in their totality to maintain the position of the eye in the orbit. The conjugated optomotor reflexes will strive in virtue of their origin to maintain the position of the eye with respect to the external world. Therefore the conjugated reflexes will react in particular to displacements of the image of the environment over the retina, and it is very probable that for orientation in the external world (adjusting and directional reflexes) the conjugated reflexes are of decisive importance.

The monocular optomotor reflexes make it possible also for opposed movements of the eyes (such as fusion movements) to be executed. But in the long run the monocular reflexes do not remain purely monocular. Points in the retinae of the 2 eyes that are situated similarly in, as regards direction and distance, with respect to the fovea repeatedly receive the same stimuli from the external world; as a result there develops a binocular junction of the cortical representations of these retinal elements, so that when a given retinal element is stimulated not only its own cortical representation but also that of one of the corresponding retinal elements sends out a motor impulse. In man the semidecussation provides a preformed anatomical substratum for such a cortical junction.

This can only strengthen an impulse to conjugated movement of the eyes. The impulse to originally monocular movement, however, thus becomes an impulse to binocular movement. Nevertheless, the cortical junction in response to the repeated demand for opposed eye movements to ensure good binocular adjustment is not a point to point junction. Each representation of a retinal element of the one eye acquires connections to a group of representations of retinal elements of the other eye. This finds expression in Panum's perception circle. Owing to the fact that this junction extends over a wider area it is possible for 2 monocular adduction impulses emanating from a single point in the external world to unite to form a single convergence impulse, which is of great importance for stereoscopic vision. As soon as a single convergence impulse is produced by a point in the external world, this point is seen single, irrespective of whether the convergence movement is executed or not.

With the eye open the monocular optomotor reflexes are continually in action as the retinal elements are continually being stimulated; they thus contribute to the optical gaze tonus. To this component of the optical gaze tonus we have given the

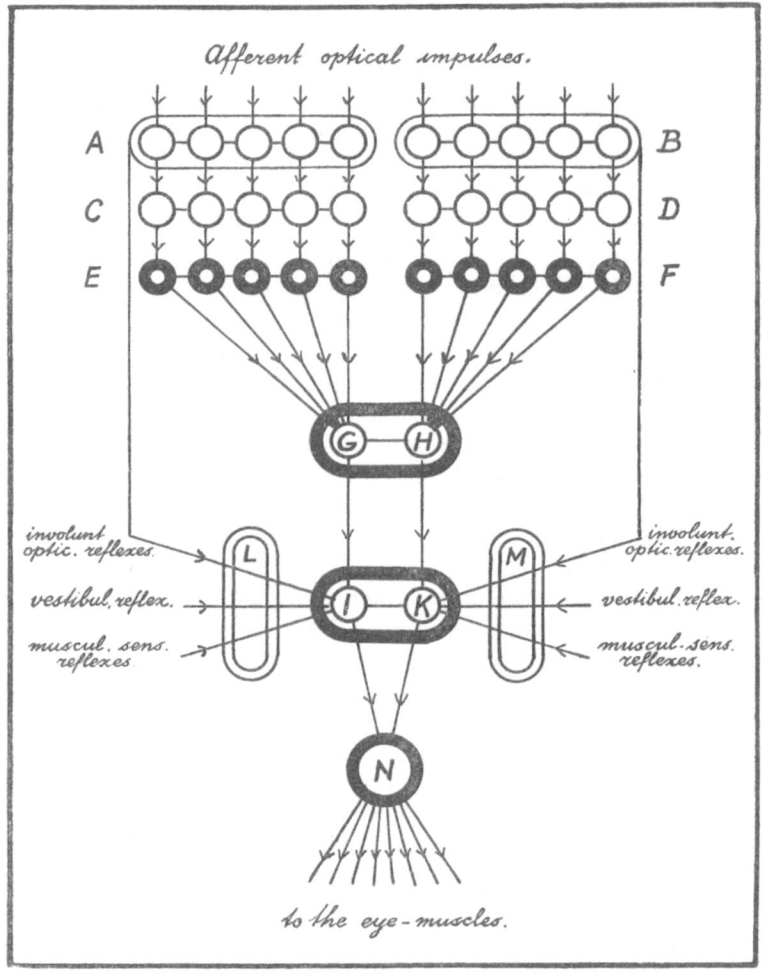

A graphic and schematic representation of the physiological correlate of optical localization;

A, B: cortical representations of the retinal elements in the left and right cerebral cortex

C, D: region in which activation and inhibition take place via associative tracts

E, F: beginning of the centrifugal cortical impulses to voluntary eye movements

G, H: higher gaze centres for voluntary eye movements

I, K: peripheral relay centres for voluntary and involuntary oculo-
       motor impulses

L. M: resultant of the involuntary oculomotor reflexes (reflex gaze
       tonus)

N: all ocular muscle nuclei

The tensions between the parts of EF and the resultant of GH determine the localization of light stimuli with respect to the fixation point

The tension between the resultant of GH and the resultant of LM determines the localization of the fixation point with respect to the optical
median plane

The resultant of the tensions in L and M determines the position of
the optical median plane

The resultant of the tensions in I and K determines the direction of gaze

name *light tonus*. The light tonus in its turn is the resultant of
a very large number of components which are provided by the
different retinal elements. These stimuli, of course, are subject
to fluctuations, so that the light tonus is not constant: it is lower
in the dark and higher in the light. The mutual relationships of
the stimuli from the different parts of the retina can also vary;
on the whole, however, the light tonus will help to maintain a
certain neutral position of the eye in the orbit.

With the conjugated optomotor reflexes matters are different.
These tend to keep the eyes at rest with respect to the environment (as do the vestibular reflexes). Every displacement of light
stimuli over the retina brings these optomotor reflexes into
action, for which reason we often refer to them as *fixation
reflexes*. These fixation reflexes do not originate solely from the
foveal region, as stated by Kestenbaum (1921, 1942), but from
all parts of the retina, although it is probable that those from
more central parts of the retina are more strongly developed.
As a result of the numberless reflex influences which regulate
its position, the eye is never at rest. The tendency to deviation
from the position once taken up (intentional direction of gaze) is
corrected by the optical fixation reflexes, so that these reflexes
too are continually at work when the eye is open. The streams
of stimuli in different directions which are required for this
purpose will also raise the gaze tonus in these directions. To this
component of the optical gaze tonus we have given the name
fixation tonus. From these considerations it follows that both
the monocular and the conjugated optomotor reflexes help to
raise the gaze tonus, but that the fixation tonus tends to adapt

itself to every position of the eye in the orbit while the light tonus tends to bring the eye back to a given position in the orbit ('Grundstellung' or neutral position).

The gaze tonus and the optomotor stimuli which constitute the tension pattern of the retinal impulses form the basis of the physiological correlate of optical localization and hence also of the visual acuity (Roelofs 1935; Mesker 1953). Let us imagine for a moment that there are in the external world 2 points of light which are depicted at different places on the retina. The elements upon which the light impinges produce 2 different optomotor impulses which strive to adjust the eyes (or perhaps the head) in two different directions at once. Thus we have 2 physico chemical processes which are in conflict with each other and give rise to a tension. This tension we regard as the physiological correlate of the apparent distance between the 2 points of light, i.e. of the exocentric optical localization. In whatever way the eye moves, the tension between the processes evoked by the 2 points of light in the external world remains the same.

As visual acuity is nothing more nor less than relative or exocentric localization — the recognition of the different parts of an object in their relationship to each other — the visual acuity will be greater the more sharply the motor impulses forming the tension pattern are differentiated, or the more closely each retinal element is linked to its own precisely graduated motor impulse. This ideal is realized in the fovea.

The egocentric localization also has its correlate in the optomotor impulses which are evoked by the objects in the external world. If a point of light in the external world gives no optomotor impulses to the eyes and head — or more correctly if the resultant of the impulses to movements of eyes and head is equal to zero — we localize this point straight ahead of us in the median plane of the body. If this resultant is only equal to zero for the movements of the eyes, the object is localized in the median plane of the head.

The question of whether a point of light will or will not evoke an optomotor impulse with a given position of the eye is partly dependent on the position which is imposed on the eye by the gaze tonus. If the eye is exactly in the physiological resting position, a point of light that forms an image at the fovea will be localized in the median plane of the head. Every point that has its image on the retina to the left or right of the fovea will give an impulse for right or left turning to the eye and will thus be localized to the right or left of the median plane. It is the tension between the existing gaze tonus and the above-mentioned optomotor impulses that determines the egocentric localization of these points. All the points in the external world which on fixation are localized in the median plane of the head form together the optical median plane.

If the eye is not in the position that is dictated to it by the gaze tonus, the image of the point of light formed at the fovea will have to evoke a continuous stream of motor impulses in order to maintain its position on the fovea and to resist the gaze tonus which is striving to bring the eye back to the physiological resting position (neutral position). Such a point, thus, is not localized in the median plane of the head. Once again it is the tension between the gaze tonus and the motor impulses evoked by the point of light that determines the optical egocentric localization.

It is easy to see that when the eye is not in the physiological resting position there must be points of light which, when depicted on the retina, give rise to optomotor impulses which are in complete harmony, as regards excursion and direction, with the existing gaze tonus. Such points are now localized in the median plane of the head, there is then no tension between these optomotor impulses and the gaze tonus. What is true of the exocentric and egocentric directional localization is also true of the depth localization. Apart from a number of secondary characteristics, the perception of a difference in depth between two points in the external world is determined by the tension between the convergence impulses evoked by each of these two points. There is a remarkable agreement between the smallest perceptible difference in direction and the smallest perceptible difference in depth (binocular parallax). For the egocentric depth or distance perception we are also dependent on the tension between the existing convergence tonus and the convergence impulse evoked by the point of light — here again leaving many secondary characteristics out of consideration.

It is clear that the exocentric or relative localization, both for direction and for depth, must be much finer than the egocentric or absolute localization. In exocentric localization the tension between the optomotor impulses evoked by the different points of light in the external world varies hardly at all; in egocentric localization on the other hand we have as basis the continually varying gaze tonus.

We have discussed optical localization at rather more length because this function is also greatly dependent on the development of the optomotor reflexes. For a normal development of the cortical binocular junction, fusion amplitude, binocular depth perception and visual acuity it is necessary that the monocular optomotor reflexes attain full development. For orientation in the external world and for the adjustment and directional reflexes a normal development of the primarily conjugated reflexes is undoubtedly also of great importance.

If the above-mentioned functions are disturbed and if no other causes of the disturbance can be found, we are certainly justified in thinking of some kind of disturbance in the development of the optomotor reflexes, the more so if we also find other anomalies which are attributable to a disturbance in the development of these reflexes. We shall now discuss some of these anomalies.

We consider strabismus (Keiner 1951) to be one of the chief anomalies resulting from a disturbance in the development of the monocular optomotor reflexes. The commonest type is strabismus convergens. In the majority of cases the convergent squint becomes manifest at an age when cortical binocular junction almost certainly does not yet exist, so that there can still be no question of a true convergence reaction. Keiner did in fact succeed in evoking dissociated eye movements in these very young squinting children, by moving a light in the field of vision of one of the eyes. This investigation provided evidence of a predominance of the adduction innervation over the abduction innervation — i.e. it showed that stronger impulses originate from the temporal halves of the retinae than from the nasal halves. If this is to lead to a convergent squint, then of course only the predominating monocular adduction reflexes can be concerned. If at a slightly older age a certain association between accommodation and convergence has become established, an existing tendency to strabismus convergens in hypermetropes can easily become a manifest strabismus convergens. The predominant adduction reflexes give an excess of adduction tonus, which is responsible for the strabismus convergens. The optical fixation reflexes, which are of conjugated nature, cannot do anything about this. They will indeed come into action with monocular fixation to compensate for the predominance of adduction tonus, and in this way it is to be expected that in monocular vision the monocular adduction tonus and the binocular fixation tonus will more or less balance each other. The binocular fixation tonus is then, in monocular vision with strabismus, different for the 2 eyes and by compensating the excess of adduction tonus it has the effect that the resulting gaze tonus directs the open eye more or less straight ahead and thus makes a correct localization possible.

In our opinion, thus, strabismus convergens originates from an excess of monocular adduction tonus, but there still remains the question of whether a strabismus divergens ever results from a predominance of monocular abduction tonus. We do not wish to deny this possibility, but we have never been able to observe a predominance of monocular abduction innervation in dissociated eye movements. The anatomical resting position of the eyes is divergent; therefore an excess of adduction tonus is always necessary to keep the eyes in the parallel position. In absence of this excess, and also with a very low light tonus, a divergent squint might occur.

Another anomaly due to a disturbance in the development of the monocular optomotor reflexes is alternating hyperphoria (Crone 1952). We have already seen that the optomotor reflexes from the nasal halves of the retinae tend to lag somewhat behind those from the temporal halves. There are also phenomena which indicate that the optomotor reflexes from the lower halves of the retinae tend to lag somewhat behind those from the upper halves. Combining these statements we see that the optomotor reflexes from the nasal lower quadrants of the retinae are in the least favoured position. Crone succeeded in giving a satisfactory explanation of all phenomena belonging to the syndrome of alternating hyperphoria on the assumption that in particular the monocular optomotor reflexes — but also the conjugated optomotor reflexes — from the nasal lower quadrants of the retinae had lagged behind in development. One of the phenomena of alternating hyperphoria to be explained in this way was latent nystagmus.

The majority of patients with alternating hyperphoria have strabismus convergens; only a few have strabismus divergens. In the latter case we are obliged to assume that the optomotor reflexes from the temporal retinal halves are also insufficiently developed.

The tension pattern to which we have already referred several times furnishes an optical gaze tonus which is made up of light tonus and fixation tonus. A very low light tonus diminishes the optical gaze tonus and we may therefore ask whether a very low light tonus can also lead to pendular nystagmus. A normal person does not get nystagmus even after a long sojourn in darkness; from this we might conclude that the optical reflex tonus is not necessary to keep the eyes at rest. This conclusion is probably premature, as in cases of congenital blindness the eyes are very rarely at rest (nystagmus of the blind). This apparent contradiction can be resolved if we assume that the optomotor reflexes also influence the development of the non-optical gaze tonus. The cortical optical reflexes, grafted onto subcortical reflex paths, probably act in such a way that they activate or inhibit the stimuli travelling along these subcortical pathways; in this way it becomes conceivable that the optomotor reflexes can also 'educate' and increase the non optical gaze tonus. If the light tonus is low but the optical fixation reflexes and the optical fixation tonus are well developed, the latter will maintain the position of the images on the retina and prevent nystagmus. The fixation reflexes, however, will not be able to act with the necessary speed unless there is a sharply differentiated graduation of the optomotor impulses from the different retinal elements. This last is absent in the amblyopic eye, which probably accounts for the nystagmus of amblyopes that is occasionally observed (Case 9). We wonder whether it would not be possible to explain miner's nystagmus in a similar way, as the foveal function is practically excluded in semidarkness.

While a deficiency of light tonus is only in very exceptional cases a cause of nystagmus, a defective development of the optical fixation reflexes and the optical fixation tonus will undoubtedly give rise very readily to nystagmus. In this connection an investigation carried out by ten Doesschate (1952; 1954) is particularly interesting. Using one eye he observed a source of light at a distance of 5 m. through a strongly refracting lens system (about 40 dioptres) mounted in a spectacle frame. By moving the frame slightly forwards or backwards he found the position in which the image of the source of light lay as precisely as possible at the centre of rotation of the eye. Under these conditions he saw a large diffusion circle that filled practically the whole aperture of the lens. In this diffusion circle he saw a large number of entoptic structures. When he now made a slight sideways movement of the eyes, the diffusion circle with its entoptic structures moved quite congruently with the eye. This means that with movement of the eye the image did not shift over the retina, so that involuntary ocular movements could not be corrected by optical fixation reflexes. If he now fixated one of the entoptic structures there occurred spontaneously — so long as he took care that the diffusion circle remained on the retina — a very regular pendular nystagmus which, when once started, could hardly be suppressed. He observed the same phenomenon in a number of other test subjects. The nonobserving eye moved in perfect conjugation with the other. This is a beautiful demonstration of the great importance of the fixation reflexes evoked by displacement of light stimuli over the retina, which prevent pendular nystagmus. But there is another point; the nystagmus did not appear until one of the entoptic structures was fixated, i.e. the nystagmic movements were also started by optomotor stimuli. This is not so surprising when one considers that every retinal image actually brings into action a number of elements with differently graduated motor impulses, which action again is probably also subject to fluctuations.

After-images are also due to a constellation of stimuli that cannot shift over the retina. Ten Doesschate did in fact also succeed in evoking a pendular nystagmus of this kind with afterimages, this having the same frequency but a smaller amplitude. Individually there were differences in frequency.

All this very strongly supports the view that nystagmus, in so far as it is of optical origin, is due chiefly to a disturbance of the optical fixation reflexes. To this view we had been led by the results of investigation of our nystagmus patients.

If the fixation reflexes which are evoked by displacement of stimuli over the retina are disturbed, both for displacement from temporal to nasal and from nasal to temporal, then a pendular nystagmus must result, as in ten Doesschate's experiments. If, however, the fixation reflexes on displacement from temporal to nasal are more disturbed (less well developed) than those on displacement from nasal to temporal, a jerking nystagmus with the fast phase in temporalward direction is to be expected. This we find in latent nystagmus. It is a remarkable fact that in such an asymmetric development of the fixation reflexes and the consequent fixation tonus, the disturbance is nearly always greater in the temporalward direction, i.e. with nasalward displacement of the contours over the retina. This we learn from the by no means rare latent nystagmus.

It seems probable that just as the optomotor reflexes from

the nasal halves of the retina develop rather more slowly than those from the temporal halves, the optical fixation reflexes to nasalward displacement of images also develop somewhat more slowly than those to temporalward image displacement.

It is difficult to say what is the cause of this. Perhaps it must be regarded as an atavistic phenomenon. If we recall ter Braak's experiments with rabbits we remember that a contour movement from behind forwards, i.e. a displacement of images over the retina from front to back gave a much stronger impulse to associated movements of the eyes than did a contour movement in the opposite direction. This can be explained biologically. If a rabbit notices something in the field of vision of, let us say, its right eye, which attracts its attention, it does not turn its eye towards this object but turns its head or body in that direction, i.e. to the right. If, now, the image of the object that has attracted attention is to remain fixed on the retina, the eye must turn to the left, i.e. forwards. In this way the movement of the eye with objects that move in the field of vision from behind forwards is undoubtedly of some use; for objects that move in the field of vision from in front backwards the utility is less obvious. If the fixation reflex to displacement of images over the retina is to be regarded as a conditioned reflex built up on the vestibular eye movement as an unconditioned reflex, it follows from the above that the conditioned fixation reflex is more effectively produced if the light stimuli have attracted attention beforehand.

In our patients with latent nystagmus we observed on several occasions that not only the temporalward optical fixation reflexes but frequently also the nasalward optical fixation reflexes were disturbed, although to a slighter degree. This was shown by the reactions to optokinetic stimulation. In about half the cases the disturbance was found to be practically equal in monocular vision with the right and with the left eye. In a number of other patients the optical fixation reflexes for the better functioning or more used eye were better developed. These cases have been placed in group IV. This group could be divided again into 2 categories: in one category only the nasalward fixation reflexes for the better eye were further developed, so that the jerking nystagmus with the use of the better eye alone was sronger than that with use of the worse eye alone. In the other category both the temporalward and the nasalward fixation reflexes for the better eye had achieved further development, so that the jerking nystagmus in monocular vision with the better eye was weaker than that in monocular vision with the worse eye.

Finally we found in a number of patients that the asymmetry in the optical fixation reflexes of the better eye was compensated by an asymmetry of the non-optical fixation reflexes; these are the cases of group V. The most nearly complete compensation was found in some of the one-eyed patients who together make up group VI. In these cases of group V and group VI the jerking nystagmus will obviously also be stronger when the good eye is covered; then the asymmetric non-optical reflexes with the asymmetric non-optical gaze tonus give rise to the jerking nystagmus.

We hope that we have succeeded in making it clear that nystagmus must be due to a too low or an asymmetric gaze tonus. If this deficiency or asymmetry is dependent on the optical reflexes, a deficiency of light tonus cannot indeed be entirely excluded, *but the disturbances in the optical fixation reflexes and in the related optical fixation tonus must be regarded in the first and most important place as the direct cause of the nystagmus.*

X. Causes of pathological pendular nystagmus.

Few people will contradict us if we say that nystagmus is due to an insufficient or abnormal gaze tonus. By the term gaze tonus we have learned to understand the excitation state of the peripheral co-ordination centres for the gaze movements and on this basis one might say that nystagmus is the consequence of a disturbance in the co-ordination centres for gaze movement. It would, however, be a sign of little scientific spirit if one were to remain content with such a statement. The investigator with an anatomical bent will immediately ask: where are these co-ordination centres? The physiologist will ask: what is the nature of the disturbance and what causes it?

As regards the localization of the peripheral gaze centres, opinions are still greatly divided. Ohm (1928, 1935) and Spiegel (1932, 1936) ascribe this rôle to the nucleus of Deiters and regard its ventro-caudal part as the relay centre for horizontal gaze movement. It is indeed quite possible that the nucleus of Deiters is the relay centre where the labyrinthine impulses and the optical impulses to conjugated movement meet and are co-ordinated, but this does not necessarily mean that it is the most peripherally situated gaze centre.

It is practically certain that the nucleus of Deiters is not fed only by labyrinthine stimuli. Unilateral destruction of a labyrinth causes a jerking nystagmus which disappears again after a relatively short time; unilateral destruction of Deiters' nucleus gives a jerking nystagmus which lasts much longer; this can only be explained on the assumption that other reflex paths besides the labyrinthine also run via this nucleus. Nevertheless, Lorente de Nó (1931), Cojazzi & Sala (1946) and Szentágothai (1943) definitely reject the idea of Deiters' nucleus as the most peripheral gaze centre, on the grounds of investigations carried out by them, and consider the substantia reticularis to be a more likely candidate. In conclusion we would also mention that van Gehuchten (1948) favours the nucleus triangularis and Muskens (1934) the nucleus commissurae posterioris as peripheral gaze centre. From these divergent opinions it is obvious that none of these investigators has been able to produce arguments sufficiently convincing to be universally accepted, so that we are obliged to wait for the results of further investigation in this difficult field.

The significance of the corpora quadrigemina and the cerebellum for gaze movement is still more uncertain. In cerebellar affections one frequently sees a jerking nystagmus which has its fast phase in the direction of the lesion and is more marked when the patient looks in that direction. Some investigators believe, however that this nystagmus in cerebellar affections is due to compression of the vestibular nuclei, the fasciculus longitudinalis posterior or the corpora quadrigemina. We do not share this opinion and we believe that destruction of certain parts of the cerebellum can indeed give rise directly te nystagmus. It does not seem impossible that the cerebellum might constitute the relay system between musculo-sensory and other stimuli. We do not propose to go into this further but we should like to quote the remark made by Cords (1923), that it is generally assumed that the cerebellum exerts a braking effect on vestibular nystagmus.

But however that may be and wherever the peripheral gaze centres may be situated, the cause of the disturbance in these centres, i.e. of the insufficient or abnormal gaze tonus, must be sought either in these centres themselves or in a deficiency or abnormal pattern of the stimuli that reach them, i.e. a disturbance somewhere in the reflex pathways. If as a result of this a given reflex system supplies insufficient or asymmetric impulses, so that the equilibrium between the co-ordination centres is upset, one is not justified in saying that all this is of vestibular origin, even though the vestibular nuclei should be the peripheral gaze centres.

The term 'central vestibular disturbance' is frequently misused to conceal our ignorance. This term is only permissible if we know that the disturbance is in the vestibular nuclei themselves or in the fibre tracts which emerge from these nuclei and transmit impulses to other co-ordination centres or to the oculomotor nuclei. It is misplaced in a case of nystagmus in which all the vestibular reactions are normal, even if there should be reason to suspect an affection of the brain-stem. In the examination of patients with nystagmus we must first strive to ascertain which reflexes are disturbed and then to find out which disturbance is to be considered as primary and which as secondary. The fact that this search may not always be successful is no reason for neglecting it.

In attempting to explain how a pendular nystagmus can be caused, we shall leave out of account those peculiar forms of nystagmus that are probably due to irradiation of stimuli, e.g. nystagmus with forceful closure of the eyelids against a resistance, owing to irradiation of the innervation from the facial nerve, or nystagmus evoked by stimulation of the cornea, the conjunctiva and some parts of the ear, owing to irradiation of the innervation from the trigeminal nerve, while hysterical nystagmus must probably also be ascribed to a certain irradiation (to the convergence centre). None of these forms of nystagmus has anything to do with optomotor reflexes.

Acquired pendular nystagmus is on the whole a rare affection — except if one includes miner's nystagmus under this heading. This is quite understandable. Pendular nystagmus is a symmetrical phenomenon and can therefore only be expected if there is a lesion just at the intersection of certain reflex paths in the median plane, or an affection with multiple and largely symmetrical localizations or an affection of certain reflex systems which are concerned in some way with the gaze movements. In this connection we must think, for instance, of syringomyelia, syringobulbia and in particular sclerosis multiplex. It is just in sclerosis multiplex that one is so readily inclined to speak of an involvement of Deiters' nucleus or a central vestibular nystagmus. As already remarked, we do not consider this permissible if the vestibular reactions are intact and we are more inclined to ally ourselves with Kestenbaum who considers the cause to be a disturbance in the tracts for the optical fixation reflexes.

Congenital pendular nystagmus is less rare than the acquired form. This is also understandable, as in such cases there is usually a disturbance in the development of a whole reflex system, so that certain reflexes for right turning and for left turning have suffered. For the sake of simplicity we shall leave vertical and rotatory nystagmus out of consideration, although congenital horizontal pendular nystagmus often shows a rotatory component.

When confronted with a case of congenital pendular nystagmus, we shall first try to ascertain which oculomotor reflexes have lagged behind in development. We can divide these reflexes into optical and non-optical. The non-optical reflexes are phylogenetically and ontogenetically older and hence less vulnerable, so that a disturbance in their development is less likely. We have, in fact, never seen a case of congenital pendular nystagmus which could be ascribed with certainty to a primary developmental disturbance of the non optical reflexes. This does not detract from the fact that secondarily the non-optical gaze tonus may be very low if there is a primary disturbance in the development of the optomotor reflexes.

We need only draw attention to the very low tonus of the ocular muscles in congenital blindness; there is certainly no-one who will ascribe this to a primary disturbance in the non-optical reflexes. Apparently the non-optical gaze tonus is 'trained' and raised during life by the optomotor reflexes. If this does not take place, Esters (1930) speaks of a secondary pathogenicity of the vestibular systems. In many cases of congenital pendular nystagmus it is more difficult to ascertain to what degree the light tonus or the fixation tonus is backward in development; in most cases both are affected but not always to an equal degree.

In congenital blindness, of course, both the light tonus and the fixation tonus are lacking. It is however, questionable in how far one may speak of a pendular nystagmus in such cases; in our opinion this is not justifiable. The movements are at one moment slow and wandering and at another moment fluttering, often resembling a jerking nystagmus in varying directions and sometimes dissociated. The stimuli to these movements are of course not optical, but the primary disturbance rendering them possible undoubtedly lies in the optical system. Similar movements are sometimes seen in the newborn and especially in babies with 'papilla grisea' (Keiner 1951). Very closely related to nystagmus of the blind is nystagmus due to darkness (Raudnitz, 1902).

In addition to the totally blind we have the almost blind patients. Some show the same phenomena as the congenitally blind, while others have a more or less coarse pendular nystagmus. In the latter group the optomotor reflexes have developed to some degree. Then we have a whole series of cases of more or less severe ocular anomalies (maculae corneae, congenital cataract, central chorioretinitis, albinism, total colour blindness). In all these cases there is unmistakably a primary disturbance in the development of the optomotor reflexes. Of our patients, Cases 8, 10, 11, 12 and 13 belonged to this group. It is difficult to say which was the more severely disturbed in these patients with ocular anomalies: the monocular optomotor reflexes with the resulting light tonus or the conjugated fixation reflexes with the resulting fixation tonus. As a properly functioning retina is necessary for the development of both systems of reflexes, it is to be expected that both systems will be defective, and this expectation was confirmed by the results of our investigation.

The situation is rather different when the optical apparatus gives good retinal images and the retinae are perfectly normal. In cases of strabismus in which the stimuli coming from one eye are suppressed in the central nervous system, this may lead to amblyopia with amblyopic nystagmus as a consequence of insufficient graduation of the optomotor impulses from the different parts of the retina and an insufficient light tonus, although the optical fixation reflexes have developed satisfactorily. There is then no marked terminal-position nystagmus and the optokinetic nystagmus is normal; the disturbance is in the monocular optomotor reflexes. Case 9 is an example of this.

Although we have no experience of miner's nystagmus, we think it likely that here also it is chiefly the monocular optomotor reflexes that become insufficient in the course of time as a consequence of the elimination of foveal function in darkness, so that we are more inclined to agree with Ohm who ascribes this form of nystagmus to a defective light tonus than with Kestenbaum (1921) and Cords (1930) who hold the fixation tonus responsible. The results of the investigations of Campbell, Harrison & Vertigen (1951), who found in patients with miner's nystagmus an insufficient convergence, defective stereoscopic vision and too prompt suppression, are in favour of the hypothesis of a disturbance chiefly in the system of the monocular optomotor reflexes. Only in more severe cases are the fixation reflexes disturbed as well. Idiopathic or hereditary nystagmus can, according to some authors, be divided further into 2 groups; in one group it is believed that the affection is carried by the sex chromosome and inherited as a recessive character, being also frequently associated with albinism, while in the other group the nystagmus is believed to be transmitted as a dominant. Waardenburg, however, pointed out in 1936 that incomplete dominance can also occur with gonosomal transmission. With X-chromosome transmission it is even stated that irregular dominant transmission occurs more than recessive; there are also a few cases known of autosomal transmission in which recessivity occurs somewhat more frequently than dominance. (Waardenburg, 1953). With hereditary nystagmus there is always a smaller or larger range within which either pendular nystagmus occurs or the eyes remain at rest (Hemmes, 1924), while with strongly lateral direction of gaze a jerking nystagmus appears. Sometimes — as in our Case 1 — this jerking nystagmus is hardly perceptible. In such a case it appears that the monocular optomotor reflexes especially have developed imperfectly. In other cases the conjugated optical fixation reflexes are clearly insufficient.

The cases of pendular nystagmus which appeared to be associated with alternating hyperphoria showed chiefly a disturbance in the development of the monocular optomotor reflexes Examples of this are Cases 5 and 6.

Finally there remained 4 patients with congenital pendular nystagmus (Cases 2, 3, 4 and 7) in whom we were unable to find any further evidence as to the origin of their nystagmus. In Cases 2 and 3 the disturbance in the conjugated optical fixation reflexes predominated strongly; in Cases 4 and 7 the monocular optomotor reflexes were, in our opinion, also severely disturbed. In view of the predominant disturbance in the optical fixation reflexes in these 4 cases, it would appear that the cause of their nystagmus differed in some way from that of the hereditary type. The possibility of anoxaemia during difficult birth, to which Anderson (1953) attaches much importance, should perhaps be considered.

The disturbance in the development of the optical fixation reflexes was shown inter alia by the marked jerking nystagmus in sideways gaze (in the pendular nystagmus group we have included also those patients who showed pendular nystagmus only in a certain direction of gaze and otherwise a jerking nystagmus with the fast phase in the direction of gaze). The direction of gaze in which pendular nystagmus was present and in which also the nystagmus was the least lively can be called the neutral position. Franceschetti (1952) determined this neutral position — which often gives rise to a torticollis nystagmica — for each eye separately in a number of patients with nystagmus. He found that in patients with parallel neutral positions for both eyes there was usually no squint, while patients who had a different neutral position for each eye usually did squint. Franceschetti believes that in this way he is on the track of a new cause of squint in cases of congenital nystagmus: If the position of the eye in which the nystagmus is at its minimum is different for each eye, he believes that the image of one eye is suppressed and that strabismus results from this. This conclusion seems to us rather premature. Franceschetti finds a correlation between dissimilar neutral positions and squint. He then concludes that these dissimilar neutral positions may lead to suppression of the image of one eye, but he neglects the possibility that suppression — which is the rule in strabismus — could just as well give rise to dissimilar neutral positions. The gaze tonus adapts itself within wide limits to the position of the eyes most used when the patient looks at something, and therefore in nystagmus patients without strabismus the neutral positions of the 2 eyes are more likely to be parallel. Franceschetti himself remarks that his theory does not hold for latent nystagmus and hence leaves this affection out of consideration as an exception to the rule. One might have expected that this exception would have suggested to him the idea of a different explanation of the observed correlation.

More important is, in our opinion, Franceschetti's observation that for all directions of gaze the frequency and the direction of the fast phase are always the same for both eyes, whereas the amplitude is not. This shows that we are always concerned with impulses to conjugated eye movements; the amplitude is governed by the resistance which has to be overcome and which is not necessarily the same for both eyes.

From the results of examination of our patients and from our theoretical considerations we have come to the following conclusions:

(1) Pendular nystagmus in most if not in all cases is due to a disturbance in the optomotor reflex path;

(2) the cause of this disturbance may lie either at the beginning of the reflex path (ocular anomalies), in higher parts of the central nervous system (developmental disturbances) or at the end of the path in question (affections of the brain stem);

(3) it is possible in certain cases to ascertain whether the pendular nystagmus is due more to a disturbance in the monocular optomotor reflexes (light tonus) or to a disturbance in the conjugated optomotor reflex pathways (fixation tonus).

XI. Pendular nystagmus and latent nystagmus.

In 1917 van der Hoeve reported in the Ann. d'. Ocul. on a patient with latent nystagmus who had formerly had a pendular nystagmus. As a special feature it may be mentioned that the transition from pendular nystagmus to latent nystagmus was also reflected in the visual acuity. In the period when pendular nystagmus was present the visual acuity in monocular vision with the right or the left eye was $^6/_{24}$ or $^6/_{18}$; after the appearance of the latent nystagmus the monocular visual acuity was reduced to $^6/_{60}$ by the coarser nystagmic movements. The binocular visual acuity, however, showed hardly any increase. We see this in the light of our opinion that the visual acuity depends on the development of the monocular optomotor reflexes and is based on the graduation of impulses from the different parts of the retina. If the pendular nystagmus disappears under the influence of a certain development of the conjugated optomotor reflexes, this does not by any means necessarily mean that the visual acuity will improve.

Those with experience in the study of cases of latent nystagmus will know that a transformation of pendular nystagmus into latent nystagmus is not uncommon. As long ago as 1918 Ohm became convinced of a close relationship between pendular and jerking nystagmus. This was one of the reasons why he regarded every type of nystagmus as an affection of the most peripherally situated relay centre for gaze movements, which he considered to be localized in the ventro caudal portion of the nucleus of Deiters. He believed that between pendular and jerking nystagmus there was no difference in kind but only in degree. Esters (1930) also believes that the cause of latent nystagmus must be sought in a lesion of the supranuclear gaze centres, because any theory must be able to account for both pendular and jerking nystagmus and also for the frequently associated fixation nystagmus. Ohm (1928) does indeed state that he observed abnormal vestibular reactions in some patients with latent nystagmus, but if it were true that latent nystagmus is practically always due to a lesion of Deiters' nucleus one would expect to find abnormal vestibular reactions in practically all cases. This does not exclude the possibility — if the tract for conjugated optical reflexes runs via Deiters' nucleus — that it may occasionally happen that a latent nystagmus is due to a bilateral affection of this nucleus. For instance, the latent nystagmus in a deaf-mute boy whose case was described by van der Hoeve in 1917 might be suggestive of a primary affection of the vestibular nuclei.

Among our own patients were 3 (Cases 20, 27 and 34) who had formerly had pendular nystagmus and now showed only latent nystagmus. Cases 27 and 34 still had pendular nystagmus in the dark — a proof that their gaze tonus was still very low.

A transition from pendular nystagmus to latent nystagmus does not, of course, occur suddenly, while the replacement of pendular by latent nystagmus may also be incomplete. Under all these conditions we can expect certain phenomena from this transition or incompleteness. It is possible, for instance, that in a case of pendular nystagmus that becomes a jerking nystagmus in sideways gaze, this jerking nystagmus may be appreciably more lively with monocular temporalward gaze than with monocular nasalward gaze. It may also happen that on optokinetic stimulation with monocular vision the reactions are more severely disturbed with temporalward contour movement than with nasalward contour movement. If we find this, then monocularly there is a predominance of nasalward fixation tonus, as is also the case in latent nystagmus. We found this in Cases 2, 6 and 10. Strangely enough we found in the same way in Cases 11, 12 and 13 that the temporalward fixation tonus predominated. These last 3 patients, however, had ocular anomalies which could be held responsible for the poor development of the optomotor reflexes, so that we do not need to consider a delayed development of reflex paths in the central nervous system.

On considering our patients with latent nystagmus we find that Cases 28 and 29 still showed, in addition to their latent nystagmus, a very clear-cut pendular nystagmus in binocular vision. This pendular nystagmus in binocular vision was still present but to a much slighter degree — sometimes very weak indeed — in Cases 17, 23, 25, 26 and 35. Case 26 is particularly important in this respect. This patient had formerly had a marked pendular nystagmus which had gradually lessened and in monocular vision had become a jerking nystagmus: at the time of examination she showed when using both eyes a fine pendular nystagmus that was also present in the dark. Finally we should like to draw attention to Case 24, who did not have a pendular nystagmus but developed one under the influence of optokinetic stimulation with binocular vision, probably because there was then no longer any optical point of support upon which the eyes could be directed. This patient's gaze tonus, thus, was very low; we think he must probably have had a pendular nystagmus in childhood.

Of the patients who had a pendular nystagmus in addition to their latent nystagmus we might also have mentioned Cases 44, 45 and 50. As these patients also had an asymmetric non-optical gaze tonus, the conditions in their cases are rather more complicated. If, however, we imagine that the excess of optical fixation tonus in the nasalward direction and the compensatory excess of non-optical fixation tonus in the temporalward direction could be abolished, then we should undoubtedly get a very lively pendular nystagmus. Thus in these cases also we may assume that the development of latent nystagmus has inhibited the pendular nystagmus.

While, as we have seen, the transition from pendular nystagmus to latent nystagmus is not such a very rare occurrence, a transition from latent nystagmus to pendular nystagmus has, as far as we know, never been observed.

From this we may conclude that pendular nystagmus appears at a lower stage of development of the optomotor reflexes than latent nystagmus. In the discussion of causes of pendular nystagmus we came to the conclusion that a low light tonus may be a contributory factor, but that in the overwhelming majority of cases the pendular nystagmus is due chiefly to an insufficient development of the conjugated optical fixation reflexes. The question we now have to answer is how latent nystagmus can emerge from it as the development of the optomotor reflexes advances.

An increase of the light tonus, such as may be expected in the first period of life, might be able in the long run to inhibit the pendular nystagmus (as we see in babies with 'papilla grisea'), but it could never account for the jerking nystagmus in monocular vision. This would be the case if the better development of the optomotor reflexes applied only to the uncrossed fibre tracts which have their origin in the temporal halves of the retinae; in binocular vision they would keep the eyes at rest by means of opposed conjugated innervations and in monocular vision they would turn the eyes in the direction of the occluded eye. For the monocular reflexes a development of this kind, giving rise to strabismus convergens, is known to occur.

But there are also serious objections to this. Uniform illumination of one eye while the other is totally occluded does not give rise to jerking nystagmus, which it would be expected to do if the light tonus evoked in the temporal part of the retina should predominate. It is true that uniform illumination of the non-fixating eye is capable of braking the jerking nystagmus of the fixating eye (Roelofs, 1928), but this is due to the resulting general elevation of the gaze tonus which will have a more or less inhibitory effect on any form of nystagmus.

The production of jerking nystagmus in patients with latent nystagmus requires contours visible to one of the eyes; it is a matter of indifference whether these contours are represented on the temporal or the nasal part of the retina of the open eye (Roelofs, 1928).

If we now enquire how contours can influence the optical gaze tonus by a reflex mechanism we find that this can only occur as a result of displacements of their images over the retina. If a further development of optical fixation reflexes is to turn a pendular nystagmus into a latent nystagmus, this can be brought about by the circumstance that the reflexes on shifting of retinal images from nasal to temporal do develop but those on shifting from temporal to nasal fail to do so. The optical fixation reflexes will always be in action as long as contours are visible, as the eyes are never completely at rest; the more the retinal image attracts attention, the more intensively the fixation reflexes make themselves felt. But the retinal images of a squinting eye, which do not penetrate to consciousness, can also suppress the latent nystagmus of the fixating eye in patients with latent nystagmus. If the central retinal images of a squinting eye also effect this suppression, this would be further proof of the intactness of the macular connections in cases of amblyopia and of the central origin of the suppression.

Our conclusion is, thus, that the replacement of pendular nystagmus by latent nystagmus is due to partial and asymmetric further development of the optical fixation reflexes, as a result of which the eyes come to rest in binocular vision, whereas in monocular vision the tendency to nasalward deviation is considerably enhanced.

XII. Causes of pathological jerking nystagmus, especially latent nystagmus.

Cases of jerking nystagmus can be classified as (A) acquired and (B) congenital.

A. *Acquired jerking nystagmus:*

This is not so rare as acquired pendular nystagmus. Many disturbances situated somewhere in the reflex pathways for gaze movements and gaze tonus are capable of giving rise to jerking nystagmus. Paresis of the ocular muscles can cause nystagmus if the patient endeavours, by a powerful gaze innervation impulse, to overcome the limitation of gaze movement imposed by the paresis. This jerking nystagmus is comparable with the physiological terminal-position nystagmus occasioned by forced sideways gazing. The adjusting impulses and the fixation reflexes which reach the peripheral gaze centres are then not deficient in the absolute sense, but they are relatively insufficient in comparison with the great resistance that they have to overcome. A gaze paresis as seen in epidemic encephalitis can sometimes also give rise to jerking nystagmus if the patient makes forcible efforts to overcome the gaze paresis. Here, however, it is questionable whether the disease involves the peripheral gaze centre itself or whether only certain tracts carrying impulses to the gaze centres are involved. In the latter case the occurrence of jerking nystagmus is much more likely, as the different systems which together provide the gaze tonus then come into conflict with each other.

Unilateral destruction of a labyrinth will give rise to jerking nystagmus towards the opposite side. The gaze tonus to the opposite side is lowered and higher demands are made on the adjusting impulses and the optical fixation reflexes for the maintenance of a direction of gaze to this side. There is then a conflict between the labyrinth tonus and the other components of the gaze tonus as a result of the asymmetry in the labyrinth tonus.

Much more resistant than the jerking nystagmus following

a lesion of the labyrinth is that following a unilateral lesion of the vestibular nuclei. This makes it probable that other impulses besides those from the labyrinth arrive in the vestibular nuclei, perhaps those of the optical fixation reflexes. With a lesion confined to the labyrinth, the vestibular nerve or the vestibular nuclei the jerking nystagmus will be towards the normal side. If, however, the reflex tracts are involved after they have left the vestibular nuclei, this will as a rule be after they have crossed and the jerking nystagmus will have its fast phase towards the affected side. In this case also, both the labyrinthine and the optical fixation reflexes may be disturbed.

If we find that the vestibular reflexes are practically normal and only slightly lowered, then a jerking nystagmus in a given direction is more likely to mean that there is a disturbance in the tracts for the optical fixation reflexes before they have reached the peripheral gaze centres or the vestibular nuclei. Here too we have a conflict between the labyrinth tonus and the optical fixation tonus, but this is now due to an asymmetry in the fixation tonus; there is then an absolute deficiency of optical fixation tonus in one direction or the other.

The various reflex tracts and relay centres for the gaze movements lie very close together in the brain stem, so that in a case of jerking nystagmus it is often very difficult to ascertain which tracts or centres are involved and where the lesion is precisely situated.

In addition to the vestibular and the optical fixation reflexes there are other non-optical and optical reflexes which may influence the gaze tonus. The course of their fibre tracts, however, is still less known. In connection with cerebellar affections, however, we would draw attention to the gaze paresis with very lively jerking nystagmus when an effort is made to look to the affected side. Jerking nystagmus has also been seen in cases of lesions of the foot of the 2nd frontal convolution where this gave rise to impairment of gaze movement in the opposite direction and the patient tried to look in this direction.

B. *Congenital jerking nystagmus.*

After this short discussion of acquired jerking nystagmus, let us now turn our attention to congenital jerking nystagmus. The forms in which jerking nystagmus appears only with sideways gaze, while there is a larger or smaller excursion range in which a pendular nystagmus is present or the eyes remain at rest, have already been mentioned in connection with pendular nystagmus (p. 192/193). We are thus left with only latent nystagmus as a congenital jerking nystagmus for further discussion.

We are fully aware of the fact that latent nystagmus does not make its appearance until a shorter or longer time after birth and is sometimes preceded by a pendular nystagmus. This in itself might be an indication of an optical origin of a disturbance which first becomes manifest under the influence of optical stimuli. But no-one will deny that latent nystagmus is due to a congenital anomaly and that no case of latent nystagmus attributable to an acquired affection has ever been described.

From the results of our investigations and the considerations arising therefrom we have become convinced that latent nystagmus is due to asymmetry of the optical fixation reflexes and the resulting asymmetric gaze tonus in monocular vision, this asymmetry being such that the fixation reflexes which move the eye nasalward are more strongly developed than those which move the eye temporalward. Among all our patients we found only 3 exceptions (Cases 11, 12 and 13) to this rule, and these, moreover, showed chiefly pendular nystagmus.

All this has been explained at length and requires no further discussion. We may enquire, however, whether this deficit in the temporal direction or excess in the nasal direction is primary or secondary.

It is not inconceivable that this asymmetry might develop as a kind of adaptation in the presence of some other disturbance. Roelofs did indeed think in 1928 that the primary disturbance consisted in a poor development of the non-optical reflexes. Since that time this opinion has been completely changed. In the first place, with an adaptation one may expect an asymmetry, but then one in which the fixation tonus in the one direction is rather higher and in the other direction rather lower than normal. In our patients, however, it was very often found, especially in the reactions to optokinetic stimulation, that the fixation reflexes were below normal in development not only for the temporalward but also for the nasalward movement. In the second place, an adaptation by asymmetric development from the 2 eyes can be of use in binocular vision (abolition of pendular nystagmus), but for patients who have been obliged from earliest infancy to rely on one eye only, it would not appear to be of any use except if there were a primary asymmetry of the non-optical gaze tonus. In our one-eyed patients, however, we believe that we have succeeded in demonstrating that the asymmetry of their non-optical gaze tonus was secondary. Moreover, there is no suggestion of an asymmetry in the non-optical gaze tonus in any of the cases of latent nystagmus with 2 good eyes. In latent nystagmus, thus, there exists a correlation between the use of both eyes and a symmetrical non-optical gaze tonus and a correlation between the exclusive use of one eye and an asymmetric non-optical gaze tonus; this shows that the asymmetric non-optical gaze tonus cannot be the cause of latent nystagmus. We are therefore convinced that *latent nystagmus must be regarded as a primary disturbance in the development of the optical fixation reflexes.*

Having come gradually to this conviction we find ourselves confronted with the following new questions: (1) Whereabouts in the optical reflex paths is the primary disturbance localized? (2) What is the nature of this disturbance?

In the present state of knowledge it is not possible to answer these questions fully, but we should like to offer a few remarks. Ocular muscle pareses can be excluded, as can also a true paresis of gaze due to lesions of the peripheral gaze centres. What is a possibility is a disturbance in the tracts which run to these peripheral gaze centres. Ohm adheres firmly to the idea that latent nystagmus is due to an affection of the nucleus of Deiters. As recently as 1942 he stated that in his opinion the nasal half of the retina sends its motor impulses via the occipital lobe of the other side to the nucleus of Deiters on the same side and that the temporal half of the retina sends its motor impulses via the occipital lobe of the same side to the nucleus of Deiters on the other side. He believes that Deiters' nucleus gives impulses to movement of the eyes to the homolateral side. In this way the stimuli from the nasal half of the retina would turn the eyes temporalwards and those from the temporal half nasalwards. Since the nasal half of the retina predominates (the temporal half of the field of vision is 30° larger), it is assumed that the stimuli from there predominate and cause the temporalward fast phase of the latent nystagmus.

We need hardly say that this view is totally incompatible with our observations and considerations. Unilateral destruction of a labyrinth causes the eyes to deviate to the side of the lesion. A unilateral lesion of the vestibular nuclei will thus do the same, so that we can hardly imagine how stimulation of the vestibular nuclei can also produce a movement of the eyes to the homolateral side. Further, the results of all our investigations indicate that it is not the nasalward but the temporalward gaze tonus that is deficient, so that one would be more inclined to suggest that the stimuli from the temporal halves of the retinae predominate.

But apart from Ohm's explanation, in which he does really seem to attach more value to the difference in strength of stimuli transmitted from the nasal and from the temporal half of the retina to the nucleus of Deiters, it seems very improbable that a disturbance in Deiters' nucleus itself, or more correctly in both nuclei of Deiters, could cause latent nystagmus. If this were so, the same co-ordination centre would have to function defectively when one eye was used but not when the other eye was used. In our opinion this makes it quite clear that the cause must lie somewhere in the afferent reflex tracts. Ohm (1928) raises the objection that an acquired disturbance in the optical reflex path has never yet been known to cause nystagmus. To this we might reply that neither has an acquired disturbance in the vestibular reflex tracts ever been known to cause latent nystagmus. However, as we have already pointed out, an acquired disturbance cannot be compared with a developmental disturbance in a given reflex system.

If we thus assume that there is an asymmetry in the impulses transmitted via the afferent reflex tract, this again cannot be due to a difference between the impulses coming from the nasal half of the retina and those from the temporal half. We have already pointed out that nystagmus with monocular vision occurs both with contours visible in the temporal half of the field of vision and with contours visible in the nasal half. We have therefore come to the conclusion that displacement of the contours over the retina from nasal to temporal evokes stronger fixation reflexes than does a displacement from temporal to nasal. The question is now where the point of action of the disturbance in the fixation reflexes is. In our opinion the 2 most likely points are firstly the area striata and its immediate neighbourhood (area parastriata and area peristriata) and secondly the retina.

In support of the hypothesis that the site of the disturbance is in the occipital cortex it might be pointed out that in the great majority of cases latent nystagmus is accompanied by strabismus. Apart from the 2 patients who had an artificial eye, there were among 40 patients with latent nystagmus only 10

who neither had a squint nor had formerly had one. As it seems
very likely that the cause of squint must be sought in the cortical
sphere, this might also be taken to apply to latent nystagmus.

Against this we have the fact that 17 of our 42 patients with
latent nystagmus had anomalies of the eye. But although this
is quite a high proportion, it is not, in our opinion, high enough
to justify the conclusion that the site of the disturbance is in the
retina, also in cases where the eyes are otherwise quite normal.
Further, it is pointed out by Verhage (1941; 1942) that in his
cases of latent nystagmus he often found not only ocular anomalies but also very often neurological and mental disturbances
(mental deficiency, imbecility, epilepsy etc.), all phenomena
which are suggestive of cerebral lesions due to an intrauterine
affection.

Just as anomalies of the eye and more centrally situated anomalies can affect the development of the optomotor reflexes in
such a way that strabismus can easily occur, these ocular and
cerebral anomalies might also impede the development of the
optical fixation reflexes.

But why then are just those optical fixation reflexes which
move the eyes temporalwards the more severely disturbed? Van
der Hoeve (1917) expressed the opinion that these normally
always lag behind the optical fixation reflexes which turn the
eyes nasalwards and that this difference becomes manifest, in
the form of latent nystagmus, if the gaze tonus in general is
low. We hesitate to endorse this statement, as so little of such
a difference is to be found in normal people (in many people a
contour movement from left to right gives a somewhat livelier
optokinetic nystagmus than a contour movement from right to
left, (Roelofs and van der Bend, 1930). Nevertheless we do consider it probable that under normal conditions the optical fixation
reflexes to displacement in the temporalward direction over the
retina develop somewhat earlier than the optical fixation reflexes
to displacement in the nasalward direction over the retina. We
have already pointed out that in the rabbit a stimulus that moves
backwards over the retina gives nystagmus more readily than
one that moves forwards over the retina (ter Braak 1936). We
suggested a biological explanation of this. The above-mentioned
difference in development might thus be an atavistic phenomenon, or it might, from an ontogenetic point of view, be a retardation phenomenon in the sense in which this term is used by
Bolk. The delay in development of the optical fixation reflexes

and especially of the fixation reflexes which move the eye in the temporal direction might then be promoted by various causes, of which we need only mention ocular and cerebral anomalies, alternating hyperphoria etc. Prenatal and postnatal development disturbances (myelogenesis retardata, Keiner 1951) are perhaps also possible.

Turning our attention now to predisposing causes, we agree with Sorsby (1931) that latent nystagmus is not a clear-cut clinical entity (as strabismus was formerly also thought to be), but all the same it is undeniable that this nystagmus must always be due to a difference in development between the nasalward and temporalward optical fixation reflexes.

In connection with this view we have been struck by the fact that our questions as to hereditary occurrence produced so few positive answers. Verhage (1941, 1942) had the same experience. Only a few of our patients stated that other members of the family had the same affection. But we know from experience how difficult it is to obtain reliable information as to heredity from patients' statements. This applies particularly to latent nystagmus, which is often not noticed by the people with whom the patient comes in contact.

We should have liked to say something about the anatomical substratum of the affection, but unfortunately nothing is known about it. We do not even know how it comes about that impulses to movement arise when light stimuli move over the retina, whereas such impulses are practically absent when a light stimulus appears and disappears at the same place. Nevertheless, one is desirous of forming some picture of what occurs and thus we are willing to describe briefly the manner in which we imagine it, although we have not the slightest proof that this corresponds to reality. We have already explained that our opinion is that when the retina is illuminated there emerge unequal motor impulses from its different parts, these forming the physiological correlate of optical localization; that these impulses more or less balance each other and that together they build up the optical gaze tonus.

Let us now imagine 3 groups of retinal elements, a, b and c, arranged in a row in such a manner that a is the nearest to the fovea and c the most peripheral, and let us call the cortical representations of these groups A, B and C. If B now suddenly receives stronger stimuli, the existing impulses from A and C will then be inhibited, so that no movement takes place. If only

A were inhibited, and if a lay to the left of b, then the eye would be turned to the left; if only C were inhibited, and c lay to the right of b, the eye would be turned to the right. If we now assume that the inhibition is strongest in that region in which the stimuli have just disappeared (Pavlov; Sherrington), it becomes clear that the eye must turn to the left when a foveafugal shift of the stimuli occurs, i.e. with displacement of the stimuli from A via B to C, and that the eye must turn to the right with foveapetal shifting of the stimuli, i.e. when the stimuli are displaced from C via B to A. This might be a picture of what happens when images move over the retina under normal conditions.

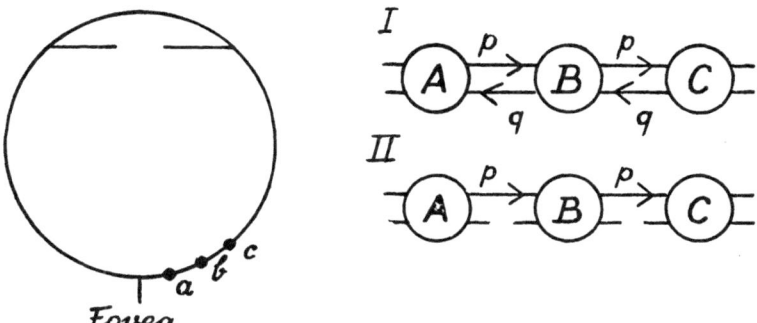

Explanation in the text.

I: p and q form the paths along which the impulse conduction and transmission (or inhibition) of neighbouring elements take place. These are here normally developed; thus there is here a normal cortical monocular junction.

II: Here there is a developmental disturbance as a result of which impulse conduction and transmission (or inhibition) can only occur in one direction. The pathways q are absent, so that stimulation of B can give an inhibition of C but not of A. The cortical monocular junction is thus partially absent.

We have assumed an inhibition of the stimuli from neighbouring elements when a light stimulus acts suddenly, as outlined above. But for this inhibition to occur it is necessary that there be a path via which the inhibitory influence can reach the cortical representations of the neighbouring retinal elements, i.e. when B receives a stronger stimulus it must have connections with both A and C in order to effect the normal inhibition. Now it is conceivable that under pathological conditions these connections might be such that although an increased stimulation of A leads to inhibition of B and an increased stimulation of B leads to inhibition of C, yet an increased stimulation of C does not lead to inhibition of B, nor an increased stimulation of B to inhibition of A. In such a case, a foveafugal displacement of a stimulus from a to c would not cause the eyes to turn to the left, but a foveapetal displacement from c to a would cause the eyes to turn to the right. Similarly one might imagine conditions under which a foveafugal displacement would give an impulse to movement of the eyes and a foveapetal displacement would not. We might sum up all this by saying that normally there is two-way traffic between neighbouring cortical representations of retinal elements of the same eye, whereas under the pathological conditions outlined above only one-way traffic has as yet been established between them. These pathological conditions would then be defects in the cortical *monocular* junction between the representations in question, analogously to the manner in which we believe (and in some cases have been able to demonstrate clinically) [1]) that defects in the cortical *binocular* junction exist in strabismus. Anatomically this might consist in disturbances of the development of the dendrites (cyto-dendro-genesis, De Crinis) or the synapses.

We thus believe that for a normal development and function of the optomotor reflex apparatus two kinds of junction between cortical representations of the retinal elements are necessary: cortical binocular junction and cortical monocular junction. If one or both of these kinds of junction is disturbed, we expect typical pathological reactions, which can be more or less accurately predicted and described, and which are at present known as clinical entities.

In this way it would be possible to gather the different affections together in a fairly simple schema in which each can be allotted its special place. A schematic presentation of this kind not only makes the field easier to survey but can improve and extend our understanding of the disturbances, while it also has certain consequence with respect to the treatment of these affections.

It still remains a remarkable fact that in latent nystagmus the motor impulses are stronger for the representations of the temporal halves of the retinae when the stimuli shift from the

---

[1]) See: New Viewpoints on the Origin of Squint by G. B. J. Keiner (M. Nijhoff, The Hague).

representations of the more central to those of the more peripheral retinal elements (foveafugal displacement), while for the representations of the nasal halves of the retinae the motor impulses (as a consequence of the inhibition) are stronger when the stimuli shift from the representations of the peripheral to those of the central retinal elements (foveapetal displacement).

Finally a few words about the fast phase of latent nystagmus. We do not believe in the existence of a kind of 'nystagmus centre' from which, constantly rhythmic impulses are supposed to emanate autonomously. We do believe that there are co-ordination centres in which reflex tracts of various origins come together. If there is a disturbance in one or more of these reflex paths, the equilibrium is upset and efforts are continually being made to restore this equilibrium via the other reflex pathways, probably by blockade of or during the refractory period of the tracts which are the site of the disturbance. In latent nystagmus it is the unequal development of the optical fixation tonus which gives rise, when only one eye is open, to the slow phase of nystagmus; the fast phase is caused chiefly by the proprioceptive reflexes (Kestenbaum's relaxation tendency: 'Entspannungstendenz') and the adjusting impulses. The adjusting impulses do not, according to Hemmes (1930), bring the eye into such a position that the point which attracts attention is depicted on the fovea, but they direct the eye in such a way that at the end of the slow phase this point has its image on the fovea. At that moment the nystagmic movement is at its slowest, so that the best visual acuity is achieved.

If we now sum up the conclusions of this last part of the discussion, these can be presented as follows:

(1) An acquired jerking nystagmus can develop as a result of a lesion somewhere in the reflex tracts which supply the innervation tonus of the eyes; this can be either in the non-optical or the optical reflex pathways.

(2) Perhaps only latent nystagmus can be regarded as a congenital jerking nystagmus.

(3) The asymmetry in the development of the optical fixation reflexes which lies at the root of latent nystagmus must be regarded not as an adaptation but as a primary disturbance.

(4) The site of the developmental disturbance in the optical fixation paths is probably in the area striata and its immediate neighbourhood in the occipital lobes.

(5) The optical fixation reflexes which move the eye nasal-wards probably take the lead in development to a small degree, also in normal individuals. The lag of the optical fixation reflexes which move the eye temporalwards may be favoured by ocular and cerebral anomalies of various kinds.

(6) No more than vague surmises can be made with respect to the anatomical substratum of the optical fixation reflexes.

(7) The fast phase of latent nystagmus is to be ascribed chiefly to proprioceptive reflexes and adjusting impulses.

SUMMARY

The preliminary notes begin with a short statement of the object of this investigation, which was chiefly to extend the knowledge of the physiology and pathology of the optomotor reflexes and to study their significance, especially in connection with the origin of nystagmus. After a few remarks on nystagmus in general and on what is to be regarded as the neutral position of the eye, latent nystagmus is discussed at rather greater length. The definition of latent nystagmus gives the opportunity to point out the highly variable picture that this form of nystagmus may present on account of many kinds of complications. Some explanations of this syndrome found in the literature are discussed.

A large part of the preliminary notes is devoted to the optomotor reflexes, how these have probably developed as conditioned reflexes (Pavlov; Zeeman) out of subcortical unconditioned reflexes and how originally monocular and conjugated optomotor reflexes can be distinguished.

The monocular reflexes, grafted onto the proprioceptive reflex paths, serve to maintain the position of the eye in the orbit; the tonic innervation supplied by them is called „light tonus". The conjugated reflexes, grafted onto the vestibular reflex paths, serve to maintain the position of the eye with respect to the external world; these are optical fixation reflexes which come into action as such when light stimuli move over the retina; they are thus postural reflexes of the eye while they can at the same time act as optical postural reflexes in the general sense (Magnus); the tonic innervation supplied by them is called „fixation tonus".

Thanks to the anatomical position of the central representations of the retinal elements of the two eyes, the foundation is laid for the formation of a cortical binocular junction between these representations both in the geniculate bodies and the occipital lobes. As a result of this the originally monocular optomotor reflexes do not remain monocular; every representation of a retinal element of one eye can form connections with the representations of a number of retinal elements of the other eye, these representations lying side by side over a certain area and forming the anatomical substratum of Panum's perception circle. The development of the convergence innervation in the more restricted sense is also related to this, having thus originated from a double adduction innervation. Under pathological conditions the cortical binocular junction may fail to develop.

The optomotor reflexes develop in such a way that different parts of the retina — and in the fovea individual elements of the retina — when stimulated send out sharply graduated impulses which together maintain a tension pattern. The sharply graded optomotor impulses form the physiological correlate of optical localization and visual acuity.

By gaze tonus is to be understood the excitation state of the most peripherally situated co-ordination centre for gaze movements. This gaze tonus is made up of various components, so that one can distinguish between a non optical and an optical gaze tonus. The non-optical gaze tonus is the resultant of a wide variety of components. The optical gaze tonus can be subdivided into light tonus and fixation tonus. The significance of quantitative and qualitative changes in the gaze tonus for the function of the eye is discussed.

Optokinetic nystagmus, which is of such great importance for the investigation of the optomotor reflexes, is discussed in detail. The differences between staring nystagmus, looking nystagmus and fixating following are briefly mentioned. The term subcortical optokinetic nystagmus is rejected as far as man is concerned. The slow phase, which is primary, is believed to manifest an increased fixation tonus resulting from a series of impulses, evoked in the pathway for the optical fixation reflex by the moving objects. The fast phase originates chiefly during a refractory period in the path of the fixation reflexes, i.e. in consequence of the abolition of the temporary blockade of all reflexes which tend to bring the eye back to the intentional direction of gaze. Attention is also paid to abnormal optokinetic reactions: optokinetic insensitivity ('optische Drehstarre') and inverse type.

The preliminary notes close with a description of the procedure of examination of the patients with nystagmus. After the history had been taken, the following were noted: refraction and visual acuity; presence of strabismus and alternating hyperphoria; ocular anomalies if any; nature and form of nystagmus in different directions of gaze; nystagmus in the dark if any; field of gaze; binocular perception; cortical binocular junction and reactions to optokinetic stimulation with each eye separately and with both eyes open.

In the chapter headed 'Case Reports' the case histories of 55 patients with nystagmus are presented; 35 of these showed typical latent nystagmus. The patients are placed in 6 separate groups.

The 1st group comprises 8 patients with pendular nystagmus and practically symmetrical gaze tonus. In all cases this abnormality could be ascribed to a deficiency in the optical gaze tonus; the fixation tonus was always insufficient but the light tonus too often left much to be desired.

The 2nd group comprises 5 patients with pendular nystagmus and an asymmetric gaze tonus. In asymmetric gaze tonus it is possible to distinguish between a binocular asymmetry, i.e. a gaze tonus that is not equal for the 2 eyes, and a monocular asymmetry, in which for each eye separately the gaze tonus is unequal for different directions of gaze. In this group there is one patient in whom only a disturbance in the development of the monocular optomotor reflexes could be found.

The 3rd group comprises 15 patients with latent nystagmus and symmetrical gaze tonus (binocular symmetry). Latent nystagmus is always due to an asymmetric development of the optical fixation reflexes for each eye separately (monocular asymmetry). The nature and degree of the disturbance in the optomotor reflexes can be best ascertained by observing the changes in the nystagmus with different directions of gaze and the reactions to optokinetic stimulation. A disturbance in the development of the monocular optomotor reflexes was also frequently found in these cases, but this did not by any means always run parallel to the disturbance in the optical fixation reflexes.

The 4th group comprises 13 patients with latent nystagmus and asymmetric optical gaze tonus (binocular asymmetry). In 7 cases the better eye showed a stronger jerking nystagmus in monocular vision; this asymmetry was ascribed to a better development of the nasalward fixation tonus from the better eye. In 6 cases the better eye showed a weaker jerking nystagmus in monocular vision; this was ascribed to a more balanced development of all the optical fixation reflexes and thus a more even fixation tonus from the better eye.

The 5th group comprises 5 patients with latent nystagmus

and an asymmetry of both the non-optical and the optical gaze tonus. These patients had jerking nystagmus in the dark as well as in daylight. The asymmetry of the non-optical gaze tonus was found to be probably secondary, having originated as a compensation of the asymmetry of the optical gaze tonus. This compensation sometimes benefited the resultant optical gaze tonus of both eyes and sometimes only that of the better and more used eye.

The 6th group comprises 7 one-eyed patients with jerking nystagmus upon occlusion of the remaining eye. The jerking nystagmus had its fast phase in the direction of the blind or absent eye. Here also there was a secondary asymmetry of the non-optical gaze tonus to compensate for the primary asymmetry in the optical gaze tonus.

For the sake of clarity the discussion is subdivided under 12 headings. Section 1 deals with the refraction and visual acuity of the nystagmus patients. The high incidence of errors of refraction makes it appear probable that the insufficient development of the optomotor reflexes and the continual movement of the eyes, as a result of which the maintenance of a correct acommodation is rendered difficult, impede the normal process of emmetropization. The visual acuity is below normal in the great majority of nystagmus patients, even when the eyes appear to be at rest. An insufficient visual acuity without discernible cause is suggestive of a defective development of the monocular optomotor reflexes, especially those from the central parts of the retina. Asymmetric development of the optomotor reflexes in the two eyes was often accompanied by unequal visual acuities.

The 2nd. section deals with strabismus and alternating hyperphoria as complications of nystagmus. The frequent occurrence of strabismus in patients with nystagmus shows that disturbances in the development of the monocular optomotor reflexes (strabismus) and of the conjugated optomotor reflexes (nystagmus) often occur together. Alternating hyperphoria was observed in about $1/3$ of the patients with latent nystagmus. Conversely, Crone found latent nystagmus in 71 % of his cases of alternating hyperphoria. Alternating hyperphoria is most probably due to a defective development of the monocular optomotor reflexes from the nasal lower quadrants of the retinae. If it is assumed that in normal individuals also the optomotor reflexes from the nasal halves of the retinae develop somewhat more slowly than those from the temporal halves, it appears conceivable that even under normal conditions the optical fixation reflexes to temporalward displacement of stimuli over the retina might develop rather earlier than those to nasalward displacement.

The 3rd section deals with the influence of the direction of gaze on nystagmus. Jerking nystagmus with sideways gaze is due to an insufficiency of the optical fixation reflexes: this insufficiency may be absolute or relative. Pendular nystagmus with sideways gaze indicates a too low tonic gaze innervation in general; this does not necessarily have to be ascribed in the first place to a disturbance of the optical fixation reflexes. An asymmetric behaviour of the eyes when the patient looks with both eyes to the right and to the left is proof of an asymmetric development of the optomotor reflexes. A dissimilar behaviour of the eye when the patient looks with one eye to the temporal and to the nasal side is proof of an unequal development of the nasalward and temporalward optical fixation tonus. A difference between the behaviour of the right eye and that of the left eye when the patient gazes sideways with one eye is proof of an unequal development of the optomotor reflexes in the two eyes. If the non-optical gaze tonus is asymmetrically developed a similar behaviour of the 2 eyes in monocular sideways gaze is impossible.

In the 4th section the reactions of the nystagmus patients to optokinetic stimulation are described. The majority of patients with pendular nystagmus show a greatly reduced sensitivity to optokinetic stimulation, both in binocular and in monocular vision. It thus appears possible that in these cases the monocular junction of the retinal representations might be defective in both directions, which would also account for the low fixation tonus. In a few cases an inverse type was seen with monocular vision, especially with temporalward movement of the contours. A normal optokinetic nystagmus both in binocular and in monocular vision may be regarded as exceptional in patients with pendular or latent nystagmus. In patients with latent nystagmus the optokinetic stimulation with both eyes open gives rise in the majority of cases to practically equal reactions for contour movement to the right and to the left. With monocular vision this is not the case. In the few cases in which this reaction is equal, the possibility of an asymmetric development of the optical and the non-optical tonic innervation must be considered. In cases of latent nystagmus it is found that a temporalward contour movement can more easily weaken an existing jerking nystagmus than a nasalward contour movement can strenghten it. As a general rule the optokinetic nystagmus has to give way to the jerking movement of latent nystagmus. An inverse type, which is found only when the fixation tonus in the direction of movement of the contours is very low, may, with a contour movement in the nasalward direction, be favoured by an excess of non-optical gaze tonus in the temporalward direction. Insensitivity to optokinetic stimuli is also indicative of a low optical fixation tonus. It was found possible, partly with the aid of optokinetic stimulation, to distinguish 3 forms of asymmetric gaze tonus. Firstly, the development of the nasalward fixation tonus may be more advanced in the better eye than in the other one. Secondly, both the temporalward and the nasalward fixation tonus from the better eye may have developed better than from the other eye. Thirdly, an asymmetric non-optical gaze tonus may have developed and thus compensated for the deficiency of temporalward optical gaze tonus of the better eye. Transitions between these 3 forms of asymmetry may be found.

Of particular interest were also the optokinetic reactions of the one-eyed patients with latent nystagmus, as they made us acquainted with an excess of non-optical fixation tonus in the temporalward direction.

The manner in which the patients with pendular nystagmus and latent nystagmus reacted to optokinetic stimuli gave clear evidence of a disturbance in the optical fixation reflexes. The type of reaction depends on (1) the severity of the disturbance in the development of the optical fixation reflexes and (2) the symmetry or asymmetry of both the optical and the non-optical reflexes. Considerations on the conjugated asymmetric tonic innervation are included in this section.

The 5th section gives an account of the ocurrence of nystagmus in the dark. This nystagmus always occurred as a conjugated movement. Where there was pendular nystagmus in daylight, this was practically always present in the dark as well. A pendular nystagmus in the dark only and not in daylight, indicates a low non-optical gaze tonus, which is probably secondary to a disturbance in the development of the optical reflexes. Jerking nystagmus in the dark is due to an asymmetric non-optical gaze tonus, which also has almost certainly developed secondarily as compensation of an asymmetric optical fixation tonus of the preferred eye. The subcortical reflexes are taken into the service of the cortical reflexes; in this way the subcortical gaze tonus is raised and regulated.

In the 6th section the binocular perception of nystagmus patients is discussed. It is proposed that the rather meaningless term 'retinal correspondence' be replaced by the expression 'cortical binocular junction'. Both in pendular and in latent nystagmus it seems that this cortical binocular junction is often absent. Evidence of its existence could be found in only about half of the 55 cases of nystagmus. Where a normal cortical binocular junction was present, it nevertheless often appeared to be far from ideal. A large number of patients showed defective depth perception, a too small fusion amplitude and a too rapid suppression of one of the retinal images. An abnormal cortical binocular junction was a great rarity among the nystagmus cases described here. Even where there was no squint, the development of the normal cortical binocular junction was defective in these patients; the development of an abnormal cortical binocular junction in cases where strabismus is present must undoubtedly be still more difficult as the anatomical conditions are less favourable. A good development of binocular perception requires a complete development of the monocular optomotor reflexes; for our nystagmus patients we must therefore suppose that the development of their monocular optomotor reflexes was more or less disturbed in the majority of cases.

In the 7th section a few remarks on optical localization in nystagmus cases are given. The results of investigation vary so widely that we considered it wiser not to draw any conclusion for the present. Perhaps we shall have to wait until we know more about optical localization in strabismus, which so often complicates nystagmus. It is possible that the position of the optical median plane will be found to depend on whether the asymmetric fixation tonus or the influence of the corrective reflexes predominates.

The 8th section is concerned with one-eyed patients with latent nystagmus; 4 cases from the literature and 7 personal cases are discussed. Occlusion of the good eye, or even simply the absence of visible contours, evoked a jerking nystagmus with the fast phase in the direction of the blind or missing eye. In these cases there was an asymmetric non-optical tonic innervation. The presence of visible contours caused the jerking nystagmus to disappear or to become transformed into a slight jerking nystagmus in the opposite direction. The asymmetries in the tonic innervation by non optical stimuli and in that by optical stimuli balance each other more or less in one-eyed patients. The predominance of the optical asymmetry with a generally too low fixation tonus and the frequently present signs of alternating hyperphoria make it probable that the optical asymmetry is primary and that the non-optical asymmetry must be regarded as a (secondary) compensation. In patients with 2 functioning eyes the asymmetric optical tonic innervations from each eye will more or less balance each other. Where this was not the case, it could be shown with a high degree of probability in some cases that here also a compensatory asymmetric non-optical tonic innervation had developed. The signs of alternating hyperphoria in the one-eyed patients showed that in addition to the disturbance in the reflexes for conjugated eye movements, as demonstrated by the latent nystagmus, there must also have been a disturbance in the development of the monocular optomotor reflexes.

The 9th section deals with gaze tonus and its significance for nystagmus. Everything that can be placed under the heading of innervation-tonus theory is once more discussed here. A continuous excitation state of the co-ordination centres for ocular movements, designated gaze tonus, is maintained all the time by numerous stimuli of very different kinds. Impulses to so-called voluntary movements of the eyes, such as adjusting and directional reflexes, are not concerned here. The reflexes which maintain the gaze tonus are divided into non-optical and optical; the optical reflexes are subdivided into originally monocular and conjugated; the former serve to maintain the position of the eyes in the orbits and the latter to maintain the position of the eyes with respect to the external world. The further development of the monocular optomotor reflexes is responsible for fusion movement, convergence, visual acuity and binocular depth perception. The further development of the conjugated optomotor reflexes serves for orientation in the external world. These conjugated optomotor reflexes — the optical fixation reflexes — come into action when light stimuli move over the retina and they give rise to the optical fixation tonus. The gaze tonus, in connection with the tension pattern that is built up by the separate impulses from the individual parts and elements of the retina, forms the basis of optical localization; this is explained further. Exocentric or relative localization is based on the tension between the optomotor impulses evoked by different visible points in the external world. Egocentric localization is based on the tension between the gaze tonus and the optomotor impulse evoked by the point to be localized in the external world. In this way it can be said that the optomotor impulses form the physiological correlate of optical localization. Disturbances in the development of the monocular optomotor reflexes must be held responsible for the occurrence of strabismus and alternating hyperphoria. Nystagmus is due to a too low or asymmetric gaze tonus; in very many cases, especially of congenital nystagmus, this disturbance in the gaze tonus is due to a disturbance in the development of the conjugated optical fixation reflexes, thus to a too low or asymmetric fixation tonus. Just as strabismus convergens is probably the consequence of a pathological retardation of the development of the monocular optomotor reflexes from the nasal hemiretinae, it is probable that latent nystagmus is the consequence of a pathological retardation in the development of the conjugated optical fixation reflexes which occur in response to nasalward displacement of stimuli over the retina. Some degree of retardation, as an atavistic phenomenon, is probably always present in normal individuals. Analogous phenomena are found in the animal kingdom and we have suggested a biological explanation for these. Whether the optical fixation reflexes will develop symmetrically or asymmetrically for the two eyes depends partly on the use of the eyes. So far as nystagmus is due to a disturbance in the optical reflexes, a deficient light tonus cannot be altogether excluded, but it is the disturbance of the optical fixation reflexes and of the resulting optical fixation tonus that must be regarded in the first and principal place as the direct cause of nystagmus.

The 10th section gives a brief survey of the causes of pendular nystagmus. It is prefaced by a few remarks as to the most peripheral gaze centre, opinions as to the localization of which are still divided. It is possible that the nucleus of Deiters forms a relay centre between the vestibular and the optomotor reflex systems, but this does not necessarily mean that it is the most peripheral co-ordination centre for gaze movements; there are other stimuli besides vestibular and optical which influence the gaze tonus.

If we leave miner's nystagmus out of consideration we can say that acquired pendular nystagmus is rather rare in comparison with congenital pendular nystagmus. Pendular nystagmus is due in most if not all cases to a disturbance in the optomotor reflex pathways. The cause of this disturbance may lie either at the beginning of the path of the reflex (ocular anomalies), at a higher level in the central nervous system (developmental disturbances) or at the end of the reflex path (affections of the brain stem). In many cases it was found possible to ascertain whether the pendular nystagmus had to be attributed more to a disturbance in the monocular optomotor reflex system (light tonus) or to a disturbance in the conjugated optomotor reflex system (fixation tonus).

In the 11th section considerations on the relationship between pendular nystagmus and latent nystagmus are presented. In addition to those patients who as children had shown a pendular nystagmus which had later disappeared completely and been replaced by latent nystagmus, we found a number of patients who already showed a suggestion of latent nystagmus along with their pendular nystagmus, while there were others who had latent nystagmus but still showed pendular nystagmus in binocular vision in daylight or in the dark. The following explanation of the relationship between pendular nystagmus and latent nystagmus was given: As long as the pendular nystagmus persists, the fixation tonus of each eye separately is deficient, both in the nasalward and in the temporalward direction. If now the optical fixation reflexes and the associated fixation tonus only achieve development in the heterolateral direction, the consequence of this is that in binocular vision the pendular nystagmus is gradually overcome, while in monocular vision a jerking nystagmus in the temporalward direction (latent nystagmus) appears.

The 12th section deals with the causes of jerking nystagmus. Acquired jerking nystagmus, which is only briefly mentioned, may occur as the result of a lesion somewhere in the reflex tracts which serve the innervation tonus of the eyes. Either the non-optical or the optical reflexes may be affected. Latent nystagmus is probably the only form that can be regarded as a congenital jerking nystagmus, as in all other cases it is nearly always possible to find a neutral position in which the eyes remain at rest or show a pendular nystagmus. The asymmetry in the development of the optical fixation reflexes which lies at the root of latent nystagmus must be regarded not as an adaptation or compensation but as a primary disturbance. The site of the disturbance in the development of the optical fixation reflexes probably lies in the area striata and its immediate neighbourhood in the occipital lobes. It is probable that a slight lag in the development of the optical fixation reflexes serving to turn the eye in the homolateral (resp. caudal) direction exists in the normal individual also; this lag might be accentuated by all kinds of ocular and cerebral anomalies. As regards the anatomical substratum for the establishment of the optical fixation reflexes, only vague surmises can be hazarded. It is suggested that cortical junctions between the monocular representations of retinal elements of each eye may be involved. The fast phase of latent nystagmus must be ascribed chiefly to proprioceptive reflexes and adjusting impulses.

REFERENCES

Anderson, J. Ringland: Causes and treatment of congenital eccentric nystagmus. Brit. Journ. of Ophthalm. Vol. 37 No. 5 p. 267—281, 1953.

Bailliart, P.: A propos d'un cas de nystagmus latent. Bull. de la Soc. d'Ophtalm, p. 66—70, 1935.

Baumeister, E.: Klinische Mitteilungen. Graefes Arch. f. Ophthalm. Bd. 19, p. 267—274, 1873.

Bárány, R.: Zur Klinik und Theorie des Eisenbahnnystagmus. Arch. f. Augenheilk. Bd. 88, p. 139—142, 1921.

Beauvieux, J.: La pseudo-atrophie optique des nouveau-nés. Revue Neurologique 1921; Paris 1926. Arch. d'Ophtalm. Vol. 7, p. 241—249, 1947.

Berg, F.: Ein Fall von latentem Nystagmus. Zeitschr. f. Augenheilk. Bd. 38, p. 164—174, 1917.

Braak, J. W. G. ter: Optokinetische Nystagmus. Ned. Tijdschr. v. Geneesk. Bd. 79, p. 1—6, 1935.

Braak, J. W. G. ter: Untersuchungen über optokinetischen Nystagmus. Arch. Neerl. Physiol. Bd. 21, p. 309—376, 1936.

Brewerton, E.: A rare form of nystagmus. Trans. Ophthalm. Soc. U. K. Vol. 23, p. 260, 1903.

Campbell, D. A., R. Harrison and J. Vertigen: Binocular vision in light adaptation in normal subjects and coal-miners. Brit. Journ. of Ophthalm. Vol. 35, p. 394—405 and 484—495, 1951.

Clarke, E.: A rare form of nystagmus. Trans. Ophthalm. Soc. U. K. Vol. 16, p. 237, 1896.

Cojazzi, L e O. Sala: Ricerche sull' interferenza tra nistagmo ottocinetico e nistagmo calorico e postrotatorio contemporaneamente provocati nell'uomo. Bull. Soc. ital. Biol. sper. Vol. 22, p. 1130—1132, 1946.

Coppez, H.: Le nystagmus. Soc. franç. d'ophtal. Rapport 1913. Vol. 30, 1913.

Cords, R.: Die Ergebnisse der neueren Nystagmusforschung. Zentralbl. ges. Ophthalm. Bd. 9, p. 369—388, 1923.

Cords, R.: Der Nystagmus. Kurz. Handb. d. Ophthalm. Bd. 3, p. 630—649, 1930.

Coutela, C.: Essai sur la coordination des mouvements des yeux á l'état normal et pathologique. Thèse de Paris, 1908.

Crinis, M. de: Anatomie der Sehrinde. Heft 64 der Monographien aus dem Gesamtgebiet der Neurologie und Psychiatrie. Julius Springer. Berlin, 1938. Wiener Klin. Wochenschr. II, 1932.

Crone, R. A.: Alternerende hyperphorie. Diss. C. V. Swets en Zeitlinger, Amsterdam, 1952.

Crone, R. A.: Alternating Hyperphoria. Brit. Journ. of Ophthalm. Vol. 38, p. 591—604, 1954.

Csapody, S. v.: Ueber Nystagmus. Arch. f. Augenheilk. Bd. 92, p. 242—259, 1923.

Doesschate, J. ten: Amblyopic nystagmus experimentally induced in non-amblyopic observers? Ophthalmologica. Vol. 124, p. 361—364. 1952.

Doesschate, J. ten: A new form of physiological nystagmus. Ophthalmologica. Vol. 127, p. 65—73, 1954.

Dorff, H.: Ueber latenten Nystagmus. Klin. Monatsbl. f. Augenheilk. Bd. 53, p. 503, 1914. Klin. Monatsbl. f. Augenheilk. Bd. 62, p. 804, 1919.

Droogleever Fortuyn, J.: Een familiair syndroom, benevens enkele opmerkingen over nystagmus in het algemeen en de snelle phase in het bijzonder. Ned. Tijdschr. v. Geneesk. Bd. 81, p. 4508—4514, 1937.

Droogleever Fortuyn, J. en H. G. van der Waals: De optokinetische nystagmus bij lijders aan haardprocessen in het centrale zenuwstelsel. Ned. Tijdschr. v. Geneesk. Bd. 84, p. 4602—4612, 1940.

Duke Elder, Sir W. Stewart: Textbook of Ophthalmology. Vol. IV. Henri Kimpyon, London, 1949.

Dupuy-Dutemps: discussion Bailliart 1935.

Esters, A.: Ueber den Nystagmus latens. Arch. f. Augenheilk. Bd. 103, p. 236—245, 1930.

Faucon, A.: Nystagmus par insuffisance des Droits externes. Journ. d'Ophtalm de Paris. Vol. 1, p. 233, 1872.

Franceschetti, A., M. Monnier and P. Dieterle: Electronystagmography in the analysis of congenital nystagmus. Transact. Ophthalm. Soc. U. K. Vol. 72, p. 515—532, 1952.

Fromaget, C. et H.: Nystagmus latent. Ann. d'Ocul. Vol. 147, p. 344—352, 1912. Ann. d'Ocul. Vol. 149, p. 241—250, 1913. Ann. d'Ocul. Vol. 153, p. 465—472, 1916.

Fromaget, C.: Reflexions sur le nystagmus latent congenital. Ann. d'Ocul. Vol. 160, p. 175—183, 1923.

Gehuchten, P. v.: Les mouvements conjugués des yeux et le système vestibulaire. Bull. schweiz. Akad. med. Wiss., p. 333—353, 1948.

Gertz, Hans: Sur le mécanisme central des mouvements des yeux. Acta med. Scandinav. Vol. 53, p. 455, 1920.

Graefe, A.: Graefe-Saemisch Handbuch. 1e Auflage, p. 227, 1880. 2e Aufl., IIer Teil, Bd. 8, Kap. II, p. 218, 1898.

Grimsdale, H. B.: A rare form of nystagmus. Transact. Ophthalm. Soc. U. K. Vol. 16, p. 238, 1896.

Hairi, H.: Nystagmus latent congenital. Rev. gén. d'ophtal. Vol. 35, p. 145, 1921.

Hemmes, G. D.: Oogbewegingen bij Nystagmus latens. Nederl. Tijdschr. v. Geneesk. 1930 II, p. 4720—4721, 1930.

Hemmes, G. D.: Zur Analyse der Augenbewegungen bei Nystagmus latens. Arch. f. Augenheilk. Bd. 103, p. 246—262, 1930.

Hemmes, G. D.: Over heredetairen Nystagmus. Diss. H. Veenman en Zonen, Wageningen, 1924.

Hoeve, J. van der: Nystagmus latent. Ann. d'Ocul. Vol. 154, p. 738—752, 1917.

Jacobs, M. W.: A case of latent nystagmus. Amer. Journ. of Ophthalm. Vol. 1, p. 171—172, 1918.

Kampherstein: Ueber die Augensymptome der multiplen Sklerose. Arch. f. Augenheilk. Bd. 49, p. 41—57, 1903.

Keiner, G. B. J.: New Viewpoints of the Origin of Squint. Thesis. M. Nijhoff, The Hague, 1951

Kestenbaum, A.: Ueber latenten Nystagmus und seine Beziehungen zur Fixation. Klin. Monatsbl. f. Augenheilk. Bd. 65, p. 426—428, 1920.

Kestenbaum, A.: Der Mechanismus des Nystagmus. Graefes Arch. f. Ophthalm. Bd. 105, p. 799—843, 1921.

Kestenbaum, A.: Zum Mechanismus des Nystagmus. Mschr. Ohrenheilk. Laryngo-Rhinol. Bd. 55, p. 844—853, 1921.

Kestenbaum, A.: Frequenz und Amplitude des Nystagmus. Graefes Arch. f. Ophthalm. Bd. 114, p. 550—582, 1924.

Kestenbaum, A.: Zum Mechanismus der Fixation. Zeitschr. f. Augenheilk. Bd. 57, p. 557—592, 1925.

Kestenbaum, A.: Zur Entwicklung der Augenbewegungen und des optokinetischen Nystagmus. Graefes Arch. f. Ophthalm. Bd. 124. p. 113—127, 1930.

Kestenbaum, A.: Zur Klinik des optokinetischen Nystagmus. Graefes Arch. f. Ophthalm. Bd. 124, p. 339—369, 1930.

Kestenbaum, A.: Visual functions in infants. Amer. Journ. of Ophthalm. Bd. 31, p. 94—95, 1948.

Kleyn, A. de: Experiments on the Quick Component Phase of Vestibular Nystagmus in the Rabbit. Versl. Akad. Amst. gewone vergad. Vol. 23, p. 1357—1364, 1921.

Lafon, Ch.: La vision des nystagmiques. Ann. d'Ocul. Vol. 151, p. 4—37, 1914.

Lafon, Ch.: Etude sur le nystagmus. Ann. d'Ocul. Vol. 157, p. 209—236 et p. 529—569, 1920.

Lagleyze, P.: Du Strabisme. Paris 1913.

Levi: Ein Fall von Nystagmus bei monocularem Sehen. Ophthalm. Klinic. ref. Nagel-Michel Jahresber. f. Ophthalm. 1901. (quoted by Kestenbaum; Arch. f. Ophthalm. 1921).

Lorente de Nó, R.: Die Labyrinthreflexe auf die Augenmuskeln nach einseitiger Labyrinthexstirpation. Berlin-Wien. 1928.

Lorente de Nó, R.: Ausgewählte Kapitel aus der vergleichenden Physiologie des Labyrinthes. Die Augenmuskelreflexe beim Kaninchen und ihre Grundlagen. Ergebn. der Physiol. Bd. 32, p. 73, 1931.

Magnus, R.: Körperstellung. Julius Springer. Berlin 1924. (Bd. 6 der Monographien aus dem Gesamtgebiet der Physiologie der Pflanzen und Tiere.)

Mesker, R. P.: De optische localisatie onder invloed van optische en houdingsfactoren. Diss. Swets en Zeitlinger. Amsterdam 1953.

Muskens, L. J. J.: Das Supra-Vestibuläre System bei den Tieren und beim Menschen, mit besonderer Berücksichtigung der Klinik der Blicklähmungen, der sogen. Stirnhirnataxie, der Zwangstellungen und der Zwangsbewegungen. N.V. Noord-Holl. Uitgevers Mij. Amsterdam, 1934.

Offergeld: Ueber nystagmusartige Zuckungen bei Gesunden. Inaug. Diss. Bonn, 1893.

Ogle, K. N.: Disparity limits of stereopsis. Arch. of Ophthalm. Vol. 48, p. 50—60, 1952.

Ogle, K. N.: On the limits of stereoscopic vision. Journ. of exper. Psychology. Vol. 44, p. 253—259, 1952.

Ohm, J.: Ueber den Einfluss des zweiäugigen Sehens auf den Nystagmus. Zeitschr. f. Augenheilk. Bd. 38, p. 269—287, 1917.

Ohm, J.: Zum 1000. Fall von Augenzittern der Bergleute. Zeitschr. f. Augenheilk. Bd. 39, p. 204—207, 1918.

Ohm, J.: Der latente Nystagmus im Stockdunkeln. Arch. f. Augenheilk. Bd. 99, p. 417—437, 1928.

Ohm, J.: Das Augenzittern der Albinos. Arch. f. Augenheilk. Bd. 103, p. 216—236, 1930.

Ohm. J.: Bemerkungen zu den Arbeiten von Prof. Dr. E. Spiegel, Philadelphia. Zeitschr. f. Hals-, Nasen-, u. Ohrenheilk. Bd. 39, p. 136— 154, 1935.

Ohm, J.: Ueber die Beziehungen zwischen angeborenem Nystagmus und willkürlich-optokinetischem Pendel- und Rucknystagmus. Graefes Arch. f. Ophthalm. Bd. 143, p. 65—68, 1941.

Ohm, J.: Der latente Nystagmus nach Verlust eines Auges. Graefes Arch. f. Ophthalm. Bd. 144, p. 617—622, 1942.

Pavlov, I. P.: Conditioned reflexes. Oxford University Press. Humphrey Milford. Oxford, 1927.

Pötzl, O. und O. Sittig: Klinische Befunde mit Hertwig-Magendiescher Augenstellung. Zeitschr. f. d. ges. Neurol u. Psychiatr. Bd. 95, p. 701 —730, 1925.

Raudnitz, R. W.: Zur Lehre von Spasmus nutans. Jahrb. Kinderheilk. Bd. 45, p. 145 u. 416, 1897.

Raudnitz, R. W.: Demonstration von experimentellem Nystagmus. Klin. Monatsbl. f. Augenheilk. Bd. 40, 2, p. 271—272, 1902.

Rochon-Duvignaud, A.: Les Yeux et la Vision des Vertébrés. Masson et Cie. Paris, 1943.

Roelofs, C. Otto: Ueber die Lokalisation mittels des Gesichtssinnes. Graefes Arch. f. Ophthalm. Bd. 113, p. 239—281, 1924.

Roelofs, C. Otto: Over fusieneiging. Ned. Tijdschr. v. Geneesk. Jg. 1926, p. 207—209, 1926.

Roelofs, C. Otto: Die Fusionsbewegung der Augen. Arch. f. Augenheilk. Bd. 97, p. 229—257, 1926.

Roelofs, C. Otto u. W. P. C. Zeeman: Optische Augenbewegung und Richtungsempfindung. Arch. f. Augenheilk. Bd. 98, p. 238—270, 1927.

Roelofs, C. Otto: Nystagmus latens. Arch. f. Augenheilk. Bd. 98, p. 401— 447, 1928.

Roelofs, C. Otto: Optische Lokalisation nach Strabismusoperation. Arch. f. Augenheilk. Bd. 99, p. 145—159, 1928.

Roelofs, C. Otto u. J. H. van der Bend: Betrachtungen und Untersuchungen über den optokinetischen Nystagmus. Arch. f. Augenheilk. Bd. 102, p. 551—625, 1930.

Roelofs, C. Otto u. H. G. van der Waals: Veränderungen der haptischen und optischen Lokalisation bei optokinetischer Reizung. Zeitschr. f. Psychol. Bd. 136, p. 5—49, 1935.

Roelofs, C. Otto: Die optische Lokalisation. Arch. f. Augenheilk. Bd. 109, p. 395—415, 1935.

Roelofs, C. Otto: Binokulare und monokulare Lokalisation. Arch. f. Augenheilk. Bd. 110, p. 330—356, 1937.

Roelofs, C. Otto: Optokinetic Nystagmus. Documenta Ophthalm. Vol. VII—VII, p. 579—650, 1954.

Sherrington, C. S.: The integrative action of the nervous system. University Press. Cambridge, 1947.

Sorsby, A.: Latent Nystagmus. Brit. Journ. of Ophthalm. Vol. 15, p. 1—18, 1931.

Spiegel, E. A. und L. Teschler: Experimentalstudien am Nervensystem. XII. Mitt. Ueber die Beziehung der Blickbahn zu den Vestibulariskernen.

Arch. f. d. ges. Physiol. Bd. 215, p. 106—119, 1926. Arch. f. d. ges. Physiol. Bd. 222, p. 359—370, 1929.

Spiegel, E. A.: Physiopathology of the voluntary and reflex innervation of ocular movements. Arch. of Ophthalm. Vol. 8, p. 738—753, 1932.

Spiegel, E. A. and N. P. Scala: The cortical innervation of ocular movements. Arch. of Ophthalm. Vol. 16, p. 967—981, 1936.

Stenvers, H. W.: Ueber die klinische Bedeutung des optischen Nystagmus für die cerebrale Diagnostik. Schweiz. Arch. f. Neurol. Bd. 14, p. 279—288, 1924.

Stenvers, H. W.: Over den optischen Nystagmus. Psychiatr. Neurol. Bladen. Jg. 1925, p. 137—152, 1925.

Straub, M.: Ueber die Aetiologie der Brechungsanomalien der Augen und den Ursprung der Emmetropie. Arch. f. Ophthalm. Bd. 70, p. 130—199, 1909.

Straub, M.: De beteekenis van den spiertonus voor de physiologie en de pathologie van het gezichtsorgaan. Ned. Tijdschr. v. Geneesk. Jg. 1910, p. 1970—1980, 1910.

Straub, M.: Een biologisch onderzoek van den tonus der spieren voerende tot een uitbreiding van de theorie der brekingstoestanden van het oog. Geneesk. Bladen. Reeks 18, p. 105—150, 1915.

Szentágothai, J.: Die zentrale Innervation der Augenbewegungen. Arch. f. Psychiatr. Bd. 116, p. 721—760, 1943.

Szentágothai, J.: Recherches expérimentales sur les voies oculogyres. La semaine des hôpitaux de Paris. Vol. 26, p. 2989—2995, 1950.

Velzeboer, C. M. J.: Bilateral cortical hemianopsia and optokinetic nystagmus. Ophthalmologica. Vol. 123, p. 187—189, 1952. Ned. Tijdschr. v. Geneesk. Bd. 96, p. 59—61, 1952.

Verhage, J. W. C.: Nystagmus latens. Ned. Tijdschr. v. Geneesk. Jg. 1941, p. 1148—1150, 1941.

Verhage, J. W. C.: Eine klinische Studie über den Nystagmus latens. Ophthalmologica. Vol. 103, p. 209—224, 1942.

Waals, H. G. van der, und C. Otto Roelofs: Ueber das Sehen von Bewegung. Zeitschr. f. Psychol. Bd. 128, p. 314—354, 1933.

Waals, H. G. van der, und C. Otto Roelofs: Veränderungen der optischen Lokalisation bei optokinetischer Reizung durch Bewegung um die sagittale Achse. Zeitschr. f. Psychol. Bd. 142, p. 200—232, 1938.

Waardenburg, P. J.: Vererbungsergebnisse und -probleme am menschlichen Auge. Zeitschr. f. indukt. Abstammungslehre. Vol. 70, p. 358—376, 1935.

Waardenburg, P. J.: Hereditaire Nystagmus. Nederl. Tijdschr. v. Geneesk. Jg. 1936, p. 5523—5528, 1936.

Waardenburg, P. J.: Zum Kapitel des ausserokularen erblichen Nystagmus. Acta genetica (Basel). Bd. 4, p. 298—312, 1953.

Wehrli, E.: Ueber 6 Fälle von latentem Nystagmus. Klin. Monatsbl. f. Augenheilk. Bd. 56, p. 444—468, 1916.

Zeeman, W. P. C.: Over de conservatieve behandeling van het scheelzien. Strabismus Symposium. Ned. Oogheelk. Gezelsch., p. 103—139, 1943.

Zeeman, W. P. C.: Biologisches zum Horopterproblem. Ophthalmologica. Vol. 118, p. 254—275, 1949.

Zeeman, W. P. C.: Conservative Treatment of Strabismus. Documenta Ophthalm. Vol. VII—VIII, p. 527—578, 1954.